THE COLONIZER'S MODEL OF THE WORLD

THE COLONIZER'S MODEL OF THE WORLD

Geographical Diffusionism and Eurocentric History

J. M. Blaut

THE GUILFORD PRESS
New York / London

© 1993 J. M. Blaut

Published by The Guilford Press
A Division of Guilford Publications, Inc.
72 Spring Street, New York, N. Y. 10012

Printed in the United States of America

This book is printed on acid-free paper.

Last digit is print number: 9 8 7 6 5 4 3 2

Library of Congress Cataloging-in-Publication Data

Blaut, James M. (James Morris)
    The colonizer's model of the world : geographical diffusionism and
eurocentric history / by J. M. Blaut.
        p.   cm.
    Includes bibliographical references and index.
    ISBN 0-89862-349-9 (hard) — ISBN 0-89862-348-0 (pbk.)
    1. History—Philosophy   I. Title.
D16.9B49   1992
901—dc20   93-22346
                                                              CIP

To Meca, Gini, and Mother

Acknowledgments

Many people contributed in many important ways to the writing of this book. Peter Taylor and Wilbur Zelinsky gave me great encouragement and wise counsel (not always heeded) during the years that I have been struggling with the issues and ideas discussed here. Among many others who contributed immensely to the book, and are happily given credit for many of the ideas it contains (the good ideas, not the errors), I wish particularly to mention Abdul Alkalimat, Samir Amin, William Denevan, Loida Figueroa, Andre Gunder Frank, William Loren Katz, José López, Kent Mathewson, Antonio Ríos-Bustamante, América Sorrentini de Blaut, and Ben Wisner. Over the years many other people have set me to thinking about the problems discussed in the book and have shown me the answers to some of these problems. Among these friends, teachers, and students, I would like to mention Chao-li Chi, Ghazi Falah, Fred Hardy, Fred Kniffen, Juan Mari Brás, Francis Mark, Sidney Mintz, Ng Hong, Doris Pizarro, Randolph Rawlins, Anselme Rémy, Waldo Rodríguez, Digna Sánchez, Howard Stanton, David Stea, and Lakshman Yapa. Peter Wissoker and Anna Brackett edited the book with patience and skill. A number of paragraphs in Chapters 3 and 4 and one in Chapter 2 are taken from an article in *Political Geography* (Blaut 1992b), and are reproduced here with the kind permission of the publisher of that journal, Butterworth-Heinemann.

Contents

CHAPTER 1

History Inside Out

THE ARGUMENT

The purpose of this book is to undermine one of the most powerful beliefs of our time concerning world history and world geography. This belief is the notion that European civilization—"The West"—has had some unique historical advantage, some special quality of race or culture or environment or mind or spirit, which gives this human community a permanent superiority over all other communities, at all times in history and down to the present.

The belief is both historical and geographical. Europeans are seen as the "makers of history." Europe eternally advances, progresses, modernizes. The rest of the world advances more sluggishly, or stagnates: it is "traditional society." Therefore, the world has a permanent geographical center and a permanent periphery: an Inside and an Outside. Inside leads, Outside lags. Inside innovates, Outside imitates.

This belief is *diffusionism*, or more precisely *Eurocentric diffusionism*. It is a theory about the way cultural processes tend to move over the surface of the world as a whole. They tend to flow out of the European sector and toward the non-European sector. This is the natural, normal, logical, and ethical flow of culture, of innovation, of human causality. Europe, eternally, is Inside. Non-Europe is Outside. Europe is the source of most diffusions; non-Europe is the recipient.[1]

Diffusionism lies at the very root of historical and geographical scholarship. Some parts of the belief have been questioned in recent years, but its most fundamental tenets remain unchallenged, and so the belief as a whole has not been uprooted or very much weakened by modern scholarship.

The most important tenet of diffusionism is the theory of "the

1

autonomous rise of Europe," sometimes (rather more grandly) called the idea of "the European Miracle." It is the idea that Europe was more advanced and more progressive than all other regions prior to 1492, prior, that is, to the beginning of the period of colonialism, the period in which Europe and non-Europe came into intense interaction. If one believes this to be the case—and most modern scholars seem to believe it to be the case—then it must follow that the economic and social modernization of Europe is fundamentally a result of Europe's *internal* qualities, not of interaction with the societies of Africa, Asia, and America after 1492. Therefore: the main building blocks of modernity must be European. Therefore: colonialism cannot have been really important for Europe's modernization. Therefore: colonialism must mean, for the Africans, Asians, and Americans, not spoliation and cultural destruction but, rather, the receipt-by-diffusion of European civilization: modernization.

This book will analyze and criticize Eurocentric diffusionism as a general body of ideas, and will try to undermine the more concrete theory of the autonomous rise of Europe. The first chapter of the book discusses the nature and history of diffusionism. Chapter 2 analyzes the theory of the autonomous rise of Europe as a body of propositions about European superiority (and "the European miracle"), then tries to disprove these propositions, one after the other. Chapter 3 discusses world history and historical geography prior to 1492, attempting to show that Europe was *not* superior to other civilizations and regions in those times. Chapter 4 argues that colonialism was the basic process after 1492, which led to the selective rise of Europe, the modernization or development of Europe (and outlying Europeanized culture areas like the United States), and the underdevelopment of Asia, Africa, and Latin America. Chapter 4 also argues that the conquest of America and thereafter the expansion of European colonialism is not to be explained in terms of any internal characteristics of Europe, but instead reflects the mundane realities of location. The chain of argument in Chapters 2, 3, and 4, as a whole, therefore, is an attempt to show that Europe did *not* have historical priority—historical superiority—over what we now call the Third World.

This may seem to be too ambitious a project for one small book. I am really making just one claim. I am asserting that a fundamental and rather explicit error has been made in our conventional past thinking about geography and history, and this error has distorted many fields of thought and action. I am going to present enough evidence to show that the belief in Eurocentric diffusionism and Europe's historical superiority or priority is not convincing: not well grounded in the facts of history and geography, although firmly grounded in Western culture. It is in a sense folklore.

THE TUNNEL OF TIME

If you had gone to school in Europe or Anglo-America 150 years ago, around the middle of the nineteenth century, you would have been taught a very curious kind of history. You would have learned, for one thing, that every important thing that ever happened to humanity happened in one part of the world, the region we will call "Greater Europe," meaning the geographical continent of Europe itself, plus (for ancient times only) an enlargement of it to the southeast, the "Bible Lands"—from North Africa to Mesopotamia—plus (for modern times only) the countries of European settlement overseas. You would have been taught that God created Man in this region: the Garden of Eden was mentioned as the starting point of human history in typical world history textbooks of the period, and these textbooks placed Eden at various points between the shores of the Mediterranean Sea and the mountains of Inner Asia.

Some of your teachers would have also claimed that only the people of this region are really human: God created the people of other places as a different, nonhuman, or rather infrahuman, species. And all of your teachers of science as well as history would have agreed that non-Europeans are not as intelligent, not as honorable, and (for the most part) not as courageous as Europeans: God made them inferior. If you had asked your teachers why Europeans are more human and more intelligent than everyone else you would perhaps have been chastised for asking such a question. You would have been told that a Christian God created and now manages the world, and it would be both silly and blasphemous to suggest that He might show the same favor to non-Europeans, non-Christians, that He does to those people who worship the True God and moreover worship Him with the proper sacrament.

If you had been studying geography as well as history back in the middle of the nineteenth century, you would indeed have learned something about the non-European world. The people living in Africa and Asia would have been depicted not only as inferior but as in some sense evil. They are the people who refused to accept God's grace and so have fallen from His favor. Africans are thus cruel savages, for whom the best possible fate is to be put to useful work, and Christianized. Chinese and Indians for some unknown reason managed to build barbaric civilizations of their own, but because they are not Europeans and not Christians, their civilizations long ago began to stagnate and regress. And, for all their splendor, these never were *real* civilizations: they are cruel "Oriental despotisms." Only Europeans know true freedom.

Ideas of course change, and if you had gone to school some 50 years later, around the turn of the century, you would have been taught a much more secular form of history, and it would have had a strongly evolutionary (though not yet Darwinian) flavor. You would have learned that the earth is very old, that life is old, and that our species itself has been around for a long time. But everything important still happened in Europe (that is, in Greater Europe). The first true man, Cro-Magnon, lived in Europe. Agriculture was invented in Greater Europe (perhaps in the continent, perhaps in the Bible Lands, Europe's self-proclaimed cultural hearth). You would have been told in world history class that the first barbaric beginnings of civilization occurred in the Bible Lands. There in the Bible Lands emerged the two Caucasian peoples who make all of history. The Semites invented cities and empires, and gave us monotheism and Christianity, but stopped at that point and then sank back into Oriental decadence. The Aryans or Indo-Europeans, freedom-loving though backward folk, built on these foundations, migrating from southeastern Europe or western Asia into and through geographical Europe, and creating the first genuinely civilized society, that of ancient Greece. Then the Romans raised civilization to its next level, and thereafter world history marched inexorably northwestward. If your school was in England you would have been told that History marched from "the Orient" (the Bible Lands) to Athens, to Rome, to feudal France, and finally to modern England—a kind of westbound Orient Express.

By now a secular picture of the geography of non-Europe had begun to be taught in European schools. Africans continued to be described as savages and Oriental societies as decadent and despotic. But important changes had taken place in the relations between Europe and non-Europe during the course of the nineteenth century, and by 1900 a particular theory about this relationship had become fixed in popular discourse and was now taught in schools as standard world geography. This was the theory (described later in this chapter) according to which non-Europeans can and do rise to a civilizational level, if not equal to that of Europeans at least near that level, under European tutelage, that is, under European colonial control.

Suppose we move forward another half century, to the history and geography taught around the end of World War II. Not much change. The first True Man is still the Cro-Magnon of Europe. Agriculture was invented in the Bible Lands; so too was barbaric civilization. True civilization still marches from Athens to Rome to Paris to London, and perhaps sets sail then for New York. Non-Europeans do not contribute much to world history, although they begin to do so as a result of

European influence. (Colonial peoples learn from their tutors; Japanese imitate successfully, and so on.) Europeans are still brighter, better, and bolder than everyone else.[2]

We can sum all of this up with an image that will prove quite useful in this book. This is the idea that the world has an Inside and an Outside. World history thus far has been, basically, the history of Inside. Outside has been, basically, irrelevant. History and historical geography as it was taught, written, and thought by Europeans down to the time of World War II, and still (as we will see) in most respects today, lies, as it were, in a tunnel of time. The walls of this tunnel are, figuratively, the spatial boundaries of Greater Europe. History is a matter of looking back or down in this European tunnel of time and trying to decide what happened where, when, and why. "Why" of course calls for connections among historical events, but only among the events that lie in the European tunnel. Outside its walls everything seems to be rockbound, timeless, changeless tradition. I will call this way of thinking "historical tunnel vision," or simply "tunnel history."

The older form of tunnel history simply ignored the non-European world: typical textbooks and historical atlases devoted very few pages to areas outside of Greater Europe (that is, Europe and countries of European settlement overseas plus, for ancient history and the Crusades, the Near East), until one came to the year 1492. Non-Europe (Africa, Asia east of the Bible Lands, Latin America, Oceania) received significant notice only as the venue of European colonial activities, and most of what was said about this region was essentially the history of empire.[3] Not only was the great bulk of attention devoted to Greater Europe in these older textbooks and historical atlases, but world history was described as flowing steadily westward with the passage of time, from the Bible Lands to eastern Mediterranean Europe, to northwestern Europe. This pattern is readily discernible if we notice the salience of places mentioned in these sources, that is, the frequency of place-name mentions for different regions at different periods. For the earliest period, place-name mentions cluster in the Bible Lands and the extreme eastern Mediterranean. For successively later periods, place-name mentions cluster farther and farther to the west and northwest, finally clustering in northwestern Europe for the period after about A.D. 1000: this is the "Orient Express" pattern to which we referred previously.

After World War II, however, history textbooks began to exhibit another, more subtle, form of tunnel history. The non-European world was now beginning to insert itself very firmly in European consciousness, in the aftermath of the war with Japan and in the midst of the intensified decolonization struggles, the Civil Rights movement in the United

States, and the like. Most newer textbooks enlarged the discussion of non-European history, and said something about the historical achievements of non-European cultures. Most textbooks gave a flavor of historicity, of evolutionary progress, to non-European history, thus departing from the older pattern, which dismissed these societies as stagnant and nonevolving. Asian societies were now described as having had an evolutionary motion, though a motion slower than that of Europe. Africa was still described as stagnant, history-less, prior to the colonial era. More salience was given to Asia. However, Africa and the Western Hemisphere still received little mention for eras prior to 1492. The pattern of place-name mentions in most (not all) texts and historical atlases still suggested a flow to the west and northwest, from the Near East to western Europe. And tunnel history dominated most textbooks in the most important matter of all, the question of "why," of *explanation.* Historical progress still came about because Europeans invented or initiated most of the crucial innovations, which only later spread out to the rest of the world. So the textbooks depicted a world in which historical causes were to be found basically inside the European tunnel of time, although historical effects were to be seen basically everywhere.[4]

Textbooks are an important window into a culture; more than just books, they are semiofficial statements of exactly what the opinion-forming elite of the culture want the educated youth of that culture to believe to be true about the past and present world.[5] As we have seen, European and Anglo-American history textbooks assert that most of the causes of historical progress occur, or originate, in the European sector of the world. Textbooks of the early and middle nineteenth century tended to give a rather openly religious grounding for this Eurocentric tunnel history. In later textbooks the Bible is no longer considered a source of historical fact, but causality seems to be rooted in an implicit theory that combines a belief that Christian peoples make history with a belief that white peoples make history, the whole becoming a theory that it is natural for Europeans to innovate and progress and for non-Europeans to remain stagnant and unchanging ("traditional"), until, like Sleeping Beauty, they are awakened by the Prince. This view still, in the main, prevails, although racism has been discarded and non-Europe is no longer considered to have been *absolutely* stagnant and traditional.

Schools are always a little behind the time when it comes to the teaching of newer topics and ideas. I wish I could report that the old notions about Inside and Outside are today just artifacts, still taught in some schools because of the usual lag between research and pedagogy, but which have been discarded by Real Scholars, those who pursue historical research and write the important and influential books on world history.

But this is not the case. In the matter that concerns us most, that of explaining the larger flows of world history, the views put forward by historical scholars today tend to be quite consonant with the theories projected in textbooks. We can set aside the fact that many of the most widely used textbooks, today as in the past, are written by prominent historical scholars. There are many complex cultural reasons why historical scholarship remains committed to Eurocentric explanations for most of the crucial developments in world history: we will discuss some of these reasons later in the present chapter and return to the question at various points throughout the book. Suffice it at this point to notice a very peculiar paradox. Historians have amassed a fine record of meticulous scholarship, and rarely indeed do we encounter prejudice or deliberate distortion in their work. Moreover, their judgments about historical causation are constrained by the same methodological rigor as we find in any other field of scholarship. It is only when we come to the larger issues of causation, matters of explaining historical progress over long periods and for larger regions, and matters of explaining profound revolutions in history, that Eurocentrism exerts an important influence on discourse, and often—as we will see—leads to the acceptance of poor theories in spite of a lack of supporting evidence.

Most European historians still maintain that most of the really crucial historical events, those that "changed history," happened in Europe, or happened because of some causal impetus from Europe. ("Europe" continues to mean "Greater Europe.") To illustrate this fact, I will list now, in historical order, a series of crucial Europe-centered propositions. All of them are accepted as true by the majority, in some cases the great majority, of European historical scholars. Some of them indeed *are* true, but that is beside the present point, which is to show that historical reasoning still focuses on Greater Europe as the perpetual fountainhead of history.

1. The Neolithic Revolution—the invention of agriculture and the beginnings of a settled way of life for humanity—occurred in the Middle East (or the Bible Lands). This view was unopposed before about 1930, and is still the majority view.
2. The second major step in cultural evolution toward modern civilization, the emergence of the earliest states, cities, organized religions, writing systems, division of labor, and the like, was taken in the Middle East.
3. The Age of Metals began in the Middle East. Ironworking was invented in the Middle East or eastern Europe and the "Iron Age" first appeared in Europe.

4. Monotheism appeared first in the Middle East.
5. Democracy was invented in Europe (in ancient Greece).
6. Likewise most of pure science, mathematics, philosophy, history, and geography.
7. Class society and class struggle emerged first in the Greco-Roman era and region.[6]
8. The Roman Empire was the first great imperial state. Romans invented bureaucracy, law, and so on.
9. The next great stage in social evolution, feudalism, was developed in Europe, with Frenchmen taking the lead.[7]
10. Europeans invented a host of technological traits in the Middle Ages which gave them superiority over non-Europeans. (On this matter there are considerable differences of opinion.).
11. Europeans invented the modern state.
12. Europeans invented capitalism.
13. Europeans, uniquely "venturesome," were the great explorers, "discoverers," etc.
14. Europeans invented industry and created the Industrial Revolution.

. . . and so on down to the present.

All of the propositions in this list are widely accepted tenets of European historical scholarship today, although (as we will see) there is scholarly dispute about some of the propositions. All of this means that you and I learned these things, perhaps in elementary school, perhaps in university, perhaps in books and newspapers. We learned that all of this is *the truth*. But is it? Clearly, some of these propositions are true. Some others are true with qualifications. But some, as I will argue in this book, are not true at all: they are artifacts of the old tunnel history, in which Outside plays no crucial role and Inside is credited with everything important and everything efficacious.

EUROCENTRIC DIFFUSIONISM

Eurocentrism

What we are talking about here is generally called, these days, "Eurocentrism."[8] This word is a label for all the beliefs that postulate past or present superiority of Europeans over non-Europeans (and over minority people of non-European descent). A strong critique of Eurocentrism is underway in all fields of social thought, and this book is certainly part of that critique.

There is, however, a problem with the word "Eurocentrism." In most discourse it is thought of as a sort of prejudice, an "attitude," and therefore something that can be eliminated from modern enlightened thought in the same way we eliminate other relic attitudes such as racism, sexism, and religious bigotry. But the really crucial part of Eurocentrism is not a matter of attitudes in the sense of values and prejudices, but rather a matter of science, and scholarship, and informed and expert opinion. To be precise, Eurocentrism includes a set of beliefs that are statements about empirical reality, statements educated and usually unprejudiced Europeans accept as true, as propositions supported by "the facts." Consider, for instance, the 14 propositions about Europe's priority in historical innovation which we listed above. Historians who accept these propositions as true would be most indignant if we described the propositions as "Eurocentric beliefs." Every historian in this category would deny emphatically that he or she holds any Eurocentric prejudices, and very few of them actually do hold such prejudices. If they assert that Europeans invented democracy, science, feudalism, capitalism, the modern nation–state, and so on, they make these assertions because they think that all of this is *fact*.

Eurocentrism, therefore, is a very complex thing. We can banish all the value meanings of this word, all the prejudices, and we still have Eurocentrism as a set of empirical beliefs.

This, in a way, is the central problem for this book. We confront statements of presumed historical and scientific fact, not prejudices and biases, and we try to show, with history and science, that the presumptions are wrong: these statements are false.

How is it that Eurocentric historical statements which are not valid—that is, not confirmed by evidence and sometimes contradicted by evidence—are able to gain acceptance in European historical thought, and thereafter survive as accepted beliefs, hardly ever questioned, for generations and even centuries? This is a crucial problem for historiography and the history of ideas. To deal with it satisfactorily would take us well beyond the scope of this book, the main concern of which is empirical history and geography. Yet the problem cannot be avoided here. Libraries are full of scholarly studies that support the Eurocentric historical positions we are rejecting and refuting in this book. The sheer quantity of this work, and the respect that is properly owed to the scholars who assembled it, makes it certain that one cannot convincingly refute these positions with the factual arguments that can be presented in one book. No matter how persuasive these arguments may be, they cannot be placed, so to speak, on one arm of a balance and be expected to outweigh all of the accumulated writings of generations of European scholars, textbook writers, journalists, publicists, and the rest, heaped up on the other arm of the balance.

So, in this book, we must make a sort of two-level argument. The main level is the empirical one: What *did* happen inside and outside of Europe in the medieval and early modern centuries, and what connections *did* take place between the two sectors in that period? At the second level, we will look at some pertinent aspects of the history of Eurocentric ideas and the social context surrounding these ideas. This will be done mainly in the present chapter, which analyzes the nature and history of diffusionist ideas and concludes with a discussion of the process of social licensing by which these ideas gain currency and hegemony, and in Chapter 2, which rather systematically examines the most important arguments for European superiority prior to 1492 and to an extent discusses their historical genealogies.

Scholars today are aware, as most were not a few decades ago, that the empirical, factual beliefs of history, geography, and social science very often gain acceptance for reasons that have little to do with evidence. Scholarly beliefs are embedded in culture, and are shaped by culture. This helps to explain the paradox that Eurocentric historical beliefs are so strangely persistent; that old myths continue to be believed in long after the rationale for their acceptance has been forgotten or rejected (as in the arguments grounded in belief in the Old Testament as literal history); that newer candidate beliefs gain acceptance without supporting evidence if they are properly Eurocentric; and that, most generally, the Eurocentric body of beliefs as a whole retains its persuasiveness and power. But there is more to the matter than this. Eurocentrism is, as I will argue at great length in this book, a *unique* set of beliefs, and uniquely *powerful*, because it is the intellectual and scholarly rationale for one of the most powerful social interests of the European elite. I will argue not only that European colonialism initiated the development of Europe (and the underdevelopment of non-Europe) in 1492, but that since then the wealth obtained from non-Europe, through colonialism in its many forms, including neocolonial forms, has been a necessary and very important basis for the continued development of Europe and the continued power of Europe's elite. For this reason, the development of a body of Eurocentric beliefs, justifying and assisting Europe's colonial activities, has been, and still is, of very great importance. Eurocentrism is quite simply the colonizer's model of the world.

Eurocentrism is the colonizer's model of the world in a very literal sense: it is not merely a set of beliefs, a bundle of beliefs. It has evolved, through time, into a very finely sculpted model, a structured whole; in fact a single theory; in fact a super theory, a general framework for many smaller theories, historical, geographical, psychological, sociological, and philosophical. This supertheory is *diffusionism*.

Diffusionism

When culture change takes place in a human community, that change can be the result of an invention that occurred *within* this community. Or it can be the result of a process in which the idea or its material effect (such as a tool, an art style, etc.) came *into* the community, having originated in some other community, in some other part of the landscape. The first sort of event is called "independent invention." The second is called "diffusion."[9] Both processes occur everywhere. So far so good. But some scholars believe that independent invention is rather uncommon, and therefore not very important in culture change in the short run and cultural evolution in the long run. These scholars believe that most humans are imitators, not inventors. Therefore diffusion, in their view, is the main mechanism for change.

The scholars who hold this view are called "diffusionists." Whenever they encounter a cultural innovation in a particular region, they are inclined to look diligently for a process of diffusion into that region from somewhere else, somewhere the trait is already in use. For instance, the fact that the blow-gun is traditionally used among some Native American peoples as well as some Old World peoples is explained by diffusionists as being the result of the diffusion of this trait from the Old World to the New: the New World people, they believe, probably did not invent the trait for themselves. Why? Because they probably were not inventive enough to do so. A larger form of this same diffusionist argument claims that the great pre-Columbian civilizations of the Americas must be, ultimately, the result of transpacific or transatlantic diffusions, because these civilizing traits (agriculture, temple architecture, writing, and so on) were found much earlier in the Old World than the New, and Native Americans probably were not inventive enough to think up these things on their own.[10] Some scholars, those who have been traditionally described as "extreme diffusionists," believe that all civilization diffused from one original place on earth: some of them think that this original source of civilization was ancient Egypt, others place it somewhere in Central Asia (for instance, the Caucasus region—which scholars used to think was the original home of the "white" or "Caucasian" race).[11]

The debates between diffusionists and their opponents have been going on for more than a century in anthropology, geography, history, and all fields concerned about long-term, large-scale cultural evolution.[12] The antidiffusionists (often called "evolutionists" or "independent-invention-ists") have tended to level two basic charges against the diffusionists:

1. Diffusionists hold much too sour a view of human ingenuity: people are in fact quite inventive and innovative, so the possibility that new culture traits will appear as a result of independent invention is actually vastly greater than diffusionists admit. So investigators should consider the possibility of independent invention in any given case, rather than assuming a priori that a diffusion process explains the situation at which one is looking.

2. Diffusionists are elitists. Every diffusion must start somewhere. An invention must take place in some one community before it begins to spread (diffuse) to others. If we accept the quite fundamental assumption that all human groups are truly human in their thinking apparatuses, and therefore broadly similar in their ability to invent and innovate—this assumption is known as "the psychic unity of mankind," a nineteenth-century label that is quaint but still in use—we would expect inventions to occur everywhere across the human landscape.[13] But most diffusionists claim that *only certain select communities are inventive.* In other words, most communities change only as a result of receiving new traits by diffusion, but some places are uniquely inventive and are the original sources of the new traits. The people of these communities are *more inventive* than are people elsewhere. The psychic unity of mankind is denied: some people, or cultures, are simply smarter than others. They are *permanent centers of invention and innovation.*

This is spatial elitism. If we make a map of this landscape, we find that it has a permanent *center* and a permanent *periphery.* For the "extreme diffusionists" the entire world was mapped out this way, at least for the pre-Christian era: the permanent center of invention and innovation was thought to be Egypt, or the "Ancient Aryan Homeland" (a mythical place located somewhere in western Asia or southeastern Europe), or the Caucasus, or some other supposed navel of the ancient world. But the charge was leveled more broadly: diffusionists as a group tended to imagine that some few places, or some one place, was the primary source from which culture spread to all the other places.

It should be evident that diffusionism is very nicely suited to the idea that the world has an *Inside* and an *Outside.* In fact, diffusionism was the most fully developed scientific (or pseudoscientific) rationale for the idea of Inside and Outside. That idea, as we saw, postulates one permanent world center for new ideas; cultural evolution everywhere else results, broadly, from the diffusion of new ideas from this permanent center. This is simply the diffusionists' map on a world scale.

We come now to a fascinating anomaly. The critics of diffusionism, in the nineteenth century and even in the twentieth century, failed entirely to grasp the full implications of their critique. None of them denied that the world has an Inside and an Outside. While criticizing the diffusionists for their rejection of the principle of the psychic unity of mankind, the antidiffusionists nonetheless believed that cultural evolution has been centered in Europe, and they therefore accepted the idea—explicitly or implicitly—that Europeans are more inventive, more innovative than everyone else.[14] This is made explicit when they write about recent centuries, and particularly when they discuss the modernizing, missionarizing effect of European colonialism. It is also implicit in their writings about ancient times. These anthropologists, archaeologists, geographers, and historians of the second half of the nineteenth century and the present century do *not* focus on the Bible Lands, and their scholarly writings do *not* display any acceptance of religious assumptions. Yet Inside and Outside are explicit. They write about the Near Eastern origins of agriculture, of urbanization, and so forth. They then move smoothly to arguments about European origins of most of the rest of civilization.

My basic argument is this: *all* scholarship is diffusionist insofar as it axiomatically accepts the Inside–Outside model, the notion that the world as a whole has one permanent center from which culture-changing ideas tend to originate, and a vast periphery that changes as a result (mainly) of diffusion from that single center. I do not argue that the formal theory of diffusionism, as it was advanced and defended by scholars in the nineteenth and early twentieth centuries, *explains* the Inside–Outside model, the mythology of Europe's permanent geographical superiority and priority. Rather, the theory developed as a result of broad social forces in Europe, and entered the world of scholarship from outside of that world—from European society. Diffusionist scholars were, in essence, elaborating and codifying this theory in the realms of scholarship within which they worked: realms like archaeology, world history, and so on.

Before we proceed further I must post a warning: the word "diffusionist" has some ambiguities, and these should not be allowed to bring confusion into the present discussion. In any given debate as to whether a novel trait in a certain place was invented by the people of that place or was received from elsewhere by diffusion, those scholars who take the latter view are supporters of a "diffusionist" position, that is, they favor the specific hypothesis of diffusion as against that of independent invention. This does not necessarily mean that they have a general propensity to favor diffusion as a causal formula. Sometimes the specific issue can be a very major problem. For instance, some scholars argue that important West African culture traits diffused across the Atlantic to America before 1492. Whether they are right or wrong in this matter, they are not arguing any sort of Eurocentric diffusionism, nor do they necessarily favor diffusion over independent invention in other contexts. But most scholars who are consistent diffusionists are also Eurocentric diffusionists.

Now I will describe Eurocentric diffusionism in somewhat formal terms as a scientific theory. That theory has changed through time, but its basic structure has remained essentially unchanged. I will describe what can be called the classical (essentially nineteenth-century) form of the theory, leaving until a later section of this chapter a discussion of the not very dissimilar modern form.

Diffusionism is grounded, as we saw, in two axioms: (1) Most human communities are uninventive. (2) A few human communities (or places, or cultures) are inventive and thus remain the permanent centers of culture change, of progress. At the global scale, this gives us a model of a world with a single center—roughly, Greater Europe—and a single periphery; an Inside and an Outside. There are a number of variants of this two-sector model. Sometimes the two sectors are treated as sharply distinct, with a definite boundary between them. (This form of the model is the familiar one. It is sometimes called the "Center–Periphery Model of the World.") Another form sees the world in a slightly different way: there is a clear and definite center, but outside of it there is gradual change, gradual decline in degree of civilization or progressiveness or innovativeness, as one moves outward into the periphery. Another variant depicts the world as divided into zones, each representing a level of modernity or civilization or development.[15] The classical division was one with three great bands: "civilization," "barbarism," and "savagery."

The basic model of diffusionism in its classical form depicts a world divided into the prime two sectors, one of which (Greater Europe, Inside) invents and progresses, the other of which (non-Europe, Outside) receives progressive innovations by diffusion from Inside. From this base, diffusionism asserts seven fundamental arguments about the two sectors and the interactions between them:

1. Europe naturally progresses and modernizes. That is, the natural state of affairs in the European sector (Inside) is to invent, innovate, change things for the better. Europe changes; Europe is "historical."

2. Non-Europe (Outside) naturally remains stagnant, unchanging,

traditional, and backward. Invention, innovation, and change are not the natural state of affairs, and not to be expected, in non-European countries. Non-Europe does not change; non-Europe is "ahistorical."

Propositions 3 and 4 explain the difference between the two sectors:

   3. The basic cause of European progress is some intellectual or spiritual factor, something characteristic of the "European mind," the "European spirit," "Western Man," etc., something that leads to creativity, imagination, invention, innovation, rationality, and a sense of honor or ethics: "European values."
   4. The reason for non-Europe's nonprogress is a lack of this same intellectual or spiritual factor. This proposition asserts, in essence, that the landscape of the non-European world is empty, or partly so, of "rationality," that is, of ideas and proper spiritual values. There are a number of variations of this proposition in classical (mainly late-nineteenth-century) diffusionism. Two are quite important:

a. For much of the non-European world, this proposition asserts an emptiness also of basic cultural institutions, and even an emptiness of people. This can be called the diffusionist *myth of emptiness*, and it has particular connection to settler colonialism (the physical movement of Europeans into non-European regions, displacing or eliminating the native inhabitants). This proposition of emptiness makes a series of claims, each layered upon the others: (i) A non-European region is empty or nearly empty of people (hence settlement by Europeans does not displace any native peoples). (ii) The region is empty of settled population: the inhabitants are mobile, nomadic, wanderers (hence European settlement violates no political sovereignty, since wanderers make no claim to territory. (iii) The cultures of this region do not possess an understanding of private property—that is, the region is empty of property rights and claims (hence colonial occupiers can freely give land to settlers since no one owns it). The final layer, applied to all of the Outside sector, is an emptiness of intellectual creativity and spiritual values, sometimes described by Europeans (as, for instance, by Max Weber) as an absence of "rationality."[16]
b. Some non-European regions, in some historical epochs, are assumed to have been "rational" in some ways and to some degree. Thus, for instance, the Middle East during biblical times was rational. China was somewhat rational for a certain period in its history.[17] Other regions, always including Africa, are unqualifiedly lacking in rationality.

Propositions 5 and 6 describe the ways Inside and Outside interact:

5. The normal, natural way that the non-European part of the world progresses, changes for the better, modernizes, and so on, is by the diffusion (or spread) of innovative, progressive ideas from Europe, which flow into it as air flows into a vacuum. This diffusion may take the form of the spread of European ideas as such, or the spread of new products in which the European ideas are concretized, or the spread (migration, settlement) of Europeans themselves, bearers of these new and innovative ideas.

Proposition 5, you will observe, is a simple justification for European colonialism. It asserts that colonialism, including settler colonialism, brings civilization to non-Europe; is in fact the natural way that the non-European world advances out of its stagnation, backwardness, traditionalism.

But under colonialism, wealth is drawn out of the non-European colonies and enriches the European colonizers. In Eurocentric diffusionism this too is seen as a normal relationship between Inside and Outside:

6. Compensating in part for the diffusion of civilizing ideas from Europe to non-Europe, is a counterdiffusion of material wealth from non-Europe to Europe, consisting of plantation products, minerals, art objects, labor, and so on. Nothing can fully compensate the Europeans for their gift of civilization to the colonies, so the exploitation of colonies and colonial peoples is morally justified. (Colonialism gives more than it receives.)

And there is still another form of interaction between Inside and Outside. It is the opposite of the diffusion of civilizing ideas from Europe to non-Europe (proposition 5):

7. Since Europe is advanced and non-Europe is backward, any ideas that diffuse into Europe must be ancient, savage, atavistic, uncivilized, evil—black magic, vampires, plagues, "the bogeyman," and the like.[18] Associated with this conception is the diffusionist myth which has been called "the theory of our contemporary ancestors." It asserts that, as we move farther and farther away from civilized Europe, we encounter people who, successively, reflect earlier and earlier epochs of history and culture. Thus the so-called "stone-age people" of the Antipodes are likened to the Paleolithic Europeans. The argument here is that diffusion works in successive waves, spreading outward, such that the farther outward we go

the farther backward we go in terms of cultural evolution. But conversely, there is the possibility that these ancient, atavistic, etc., traits will *counterdiffuse* back into the civilized core, in the form of ancient, magical, evil things like black magic, Dracula, etc.

The main oppositions between the two sectors can be shown in tabular form. The following contrast-sets are quite typical in nineteenth-century diffusionist thought:

Characteristic of Core	Characteristic of Periphery
Inventiveness	Imitativeness
Rationality, intellect	Irrationality, emotion, instinct
Abstract thought	Concrete thought
Theoretical reasoning	Empirical, practical reasoning
Mind	Body, matter
Discipline	Spontaneity
Adulthood	Childhood
Sanity	Insanity
Science	Sorcery
Progress	Stagnation

What I have described thus far is, of course, a highly simplified version of the diffusionist world model. We will add qualifications and modifications as we proceed, and in particular we will see that there are significant differences between the classical form of diffusionism and the modern form of the model.

So much for what diffusionism *is*. What does it *do*? In this book I will show in some detail how diffusionism has shaped our views of history, both European and non-European. Later in the present chapter and in Chapter 2 I will show some of the concrete influences of diffusionism on theories outside of history, some in psychology, some in geography, some in economics, some in sociology.

But this discussion will be more meaningful after a different question has been addressed. This is the question of how and why diffusionism became such a foundation theory in Western thought. To this question we now turn.

THE COLONIZER'S MODEL

Perhaps all civilizations have a somewhat ethnocentric view of themselves in relation to their neighbors, believing themselves better, brighter, and bolder than every other human community and constructing empirical theories to explain why this is the case, and to explain away embarrassments. Perhaps we would find the seeds of diffusionism in all these beliefs: the idea that progress is natural and calls for no explanation in "our" society but is unnatural or at, any rate, less impressive in "their" societies; the idea that "they" progress by borrowing from "us" and imitating "our" ideas; and so on. But this does not add up to the theory of diffusionism, nor is it very important for an understanding of diffusionism. Diffusionism as I have been discussing it is a product of modern European colonialism. It is the colonizer's model of the world.

Origins

Diffusionism became a fully formed scientific theory during the nineteenth century. The origins of the theory, however, go back to the sixteenth and seventeenth centuries in Western Europe, where a belief system was being constructed to give some coherence to the new reality of change within Europe and colonial expansion outside Europe. The conception of a two-sector world, the diffusionist distinction between Inside and Outside, emerged from a very old conception of Christendom and the Roman imperial legacy (which meant, for most of western and southern Europe, a common source of legitimacy for the political and landholding elites). But it is certainly not true that medieval Europeans saw Christendom as a sharply defined space, naturally in conflict with surrounding societies. Nor did medieval European thinkers have many illusions about the relative power, wealth, and technological prowess of Christendom as compared with Islamic and Oriental civilizations. But certainly they had some idea, not of a collective identity, but of a distinction between the lands inhabited by Christians, given divine guidance and protection for this reason, and the lands of non-Christians. Nevertheless, it was only after 1492 that Inside acquired a sharp geographical definition, less as a result of medieval ideas than of colonialism in the early modern period.

European ideas about European progressiveness, Europe's somehow inevitable progress, are also essentially postmedieval. Obviously, these ideas were being discussed during the Middle Ages, and certainly there was hope, prayer, and struggle for betterment, but medieval folk tended rather to see their society as being in a relative state of equilibrium; their religion spoke of the Fall, and of the need to accept existing conditions (and rules), while the reality of medieval life (particularly fourteenth- and fifteenth-century life) was not one of perceptible forward progress for the mass of people. But European thinkers of the sixteenth and seventeenth centuries were coming to conceive of history—their own history—as a progressive process. Real progress (or at any rate accelerating change) indeed was taking place in the communities occupied by these thinkers, and the climate of ideas was changing in a complex but close association with the rise of capitalism, the expansion of opportunity for individuals in many regions, and the like.

One of the main problems confronted by these early modern thinkers, both secular and religious, was the need to establish a belief system, an ideology, that would convince conservative sectors of the European community to accept the idea that progress is inevitable, natural, and desirable, and thus to accept changes in the legal system which would permit more rapid and widespread capital accumulation, to persuade the landowning classes to treat land as a commodity and invest their real holdings in risk enterprises, to introduce laws and practices to mobilize labor for emerging capitalist activities at home and abroad, to persuade Europeans in general to accept the painful changes being imposed on them, and so on. Equally important was the need to explain progress in ways that accorded with religion. This was done by seeing God's guidance of (European) history, and by conceptualizing progressive innovations as being products of the European mind or spirit and thus ultimately products of the Christian soul. (We will go into these matters later in this book.)

Thus emerged the conception of Inside as being naturally progressive (diffusionist proposition 1), and as being progressive because of the workings of an intellectual or spiritual force, "European rationality" (diffusionist proposition 3). By the eighteenth century it had become the practice in secular writings to discuss causality in history and philosophy without referencing God and Scriptures, but the basic model of European progress as natural, as rational, remained unchanged in its essence; became, indeed, much fortified.[19] In none of this thought was there any real suspicion that non-European civilizations might have had much to do with the earlier, medieval progress of Christian Europe, or much to do with its modern (sixteenth- to eighteenth-century) progress except in the purely passive role of provider of labor, commodities, and land for settlement, and, marginally, in some traits of technology and art. Nor was there very much awareness that colonialism and its windfalls—inflow of capital, intensification of intra-as well as extra-European trade, increase in employment opportunities in mercantile centers and in the colonial world, and much more—was an important cause of European progress. Then, as later, the European conception of its own dynamic society attributed dynamism not to external causes but to internal causes, and to God. This relatively constant blindness to the importance of colonialism, historically and even today, will claim our attention at various points in this book.

The development of a conception of Outside proceeded in a more complicated way. The sixteenth-century Spanish debates about the nature of New World Indians—Are they human? Can they receive the True Religion and, if so, can they be made slaves?—was a crucial part of the early formulation of diffusionism, because it entailed an attempt to conceptualize European expansion and explain why it was, somehow, natural, desirable, and profitable, and to conceptualize the societies that were being conquered and exploited, explaining why it was, again, natural for them to succumb and to provide Europeans with labor, land, and products.[20] European views of New World peoples were formed rather quickly because in that region the basic pattern of colonialism emerged quickly. The enterprise was immensely profitable from the very beginning, from the first great shipments of gold in the early sixteenth century (see Chapter 4). Resistance was rapidly overcome in Spanish colonies and surviving Americans were rather quickly forced to submit to colonial exploitation. (I refer here to the major centers of early colonialism, such as central Mexico, the Greater Antilles, and the Andes.) In the next century the profitability of slave plantations in Brazil and the Antilles, and the fact that African slaves could, in spite of their resistance, be forced to work and produce profit for Europeans, added further to the conception of Outside: these people were naturally inferior to Europeans, naturally less brave, less freedom loving, less rational, and so on, and progress for them depended on acceptance of European domination, hence diffusion. In sum, the New World experience, with Native Americans, Africans, mestizos, and mulattos, in areas of mining, large-scale estate agriculture, commercial plantations, and so forth, produced the kernel of the diffusionist propositions (2, 4, and 5), which assert that non-Europe naturally depends on Europe for progress and this is due to a lack of the intellectual and spiritual qualities that Europeans possessed. It also produced the diffusionist proposition (6) that European expansion is natural and leads naturally to the transfer of wealth from non-Europe to Europe.

But these activities were taking place only in the New World and the slave-trading coasts of Africa. The civilizations of Sudanic, southern, and eastern Africa were not conquered by Europeans until (in most cases) the nineteenth century; the Ottoman Empire was not only formidable throughout this period, but was indeed expanding its territorial control in southeastern Europe at the same time that Iberians were conquering the New World. And the other great empires did not begin to succumb to colonialism before the mid-eighteenth century: European activities in nearly all parts of Asia and Africa were mainly matters of trade, with dominance of long-distance maritime trade, small territorial footholds here and there on certain coasts, and the like (see Chapter 4). Thus, in the sixteenth, seventeenth, and eighteenth centuries, the diffusionist model of Outside which had been applied to Americans and to the groups of Africans transported to America as slaves could not be applied to the civilizations of the Old World.

For these Eastern Hemisphere civilizations a limited and somewhat tentative form of the diffusionist model was accepted during this period. These civilizations were indisputably rich and technically advanced. Rationality in the sense of inventiveness was clearly present, or at least had been present in the past, when the great and impressive innovations (technology, architecture, banking, and so on) had occurred. What these civilizations lacked was the moral component of rationality, because, fundamentally, they were not Christians. These were "Oriental despot-isms," societies in which there was, naturally, cruelty, lack of freedom, lack of a decent life for the common people (while the elite wallowed in decadence and sin), etc. (We discuss this concept at some length in Chapter 2.) The moral failing necessarily led to an inability of these societies to progress in the present, the era when Europe is naturally progressing, although Europeans could not deny, even though they found it most puzzling, that these despotic civilizations had, indisputably, progressed in the past. (At times the puzzle was resolved by declaring that they had progressed in pre-Christian times, and had lost the Grace of God through having refused to accept Christianity.[21]) Only in the nineteenth century, with the rapid colonization of India, Southeast Asia, interior Africa, and (as a kind of collective colony) China, did the diffusionist propositions about Outside, and about the natural relations between Inside and Outside, become generalized to all of non-Europe. It was in this late period, I think, that the final diffusionist proposition, the notion of counterdiffusion of evil and savagery and disease from Outside to Inside, become fully developed. The bogeyman (from the Malay Buginese people) came from Outside. So too did Dracula, whose homeland lay on the edge of Asia.[22]

Classical Diffusionism

The nineteenth century was the classical era for colonialism, and the era when Eurocentric diffusionism assumed what I will call its classical form. After the Napoleonic Wars colonialism expanded and intensified with remarkable rapidity. Between 1810 and 1860 or thereabouts Europeans subdued most of Asia, settled most of North America, and began the penetration of Africa. Between 1860 and the start of World War I, the rest of Asia and Africa was occupied and the profits from colonialism, the value of capital accumulated in Asia, Africa, and Latin America, along with the riches flowing from newly settled areas of European settlement, expanded enormously. In the latter part of the century the rate of growth in colonial agricultural enterprise exceeded that of industrial development on a world scale, and other forms of development, such as the mining of nonprecious metals, were becoming important for the first time.[23] There is profound disagreement as to how important all of this was for European social and intellectual evolution in this period. I will enter this debate in Chapter 4, but for the present it is sufficient to assert that the overall effect of European expansion, through colonialism in the narrow sense, through settlement, and through semicolonial economic dominance, was profound enough to create a very large intellectual model, the classical form of diffusionism.

By 1870 or thereabouts there was broad agreement among European thinkers about the basic nature and dynamics of the world. Few doubted that biological and social evolution—that is, progress—were fundamental truths, although evolutionary processes were more often explained in religious or metaphysical ways than in naturalistic ways, as in Darwin's theory.[24] It seemed clear that Europeans were naturally to experience permanent social evolution, that this had been God's or Nature's plan throughout history. Some historical thinkers described the general process in holistic terms, as the evolution of society or the state; others treated it in reductionist (in a sense, psychological) terms as an intellectual ascent, a matter of the steady advance of human reason, with capitalism, industry, and so on, treated as products of mind; but many thinkers (among them Herbert Spencer) saw no opposition between the social and intellectual models, treating progress as a kind of flowing stream, which carried with it an evolution both of society and of mind.[25] All of this was explicit with regard to Europe and Europeans including of course Anglo-Americans. Thus by the 1870s at the latest the central diffusionist proposition, the notion of natural, continuous, internally generated progress in the European (or west European) core, was very firmly in place. Its truth was no longer really questioned by mainstream thinkers.

There was, at the same time, a convergence of views about the nature and historical dynamics of the non-European world. By the middle of the nineteenth century the biblical time scale had been rather definitively rejected (though not yet in all history textbooks), and it was no longer necessary to argue that differences between Europeans and all other cultures had to have evolved in the space of a few thousand years unless they had been there from the start, unless, that is, polygenists were right and some human groups had been created separately from, and perhaps much earlier than, Adam and Eve. This gave room for wide-ranging theorization about the way cultural differences had evolved. Paralleling this change was a general rejection of the literal biblical beliefs about the original nature of human society. Culture was now (after midcentury) quite generally seen as a product of evolution from very primitive beginnings, exemplified in the notion of a primordial "stone age." (According to the Old Testament, humans had possessed advanced technology, including agriculture and the use of metals, in the days of Genesis.)

The reasons for the rapid crystallization of beliefs about non-Europeans are complex, but the most important underlying reason was the progress of colonialism. This produced two effects in particular. One was a flood of information about non-European people and places, such that, for the first time, a coherent—though highly distorted—description could be given in the European literature about non-Europeans, both civilized and "savage." The second reason was a practical, political and economic interest in proving certain things to be true, and other things untrue, about the extra-European world and its people. The two processes were tightly interconnected.

Colonialism in its various forms, direct and indirect, was an immensely profitable business and considerable sums of money were invested in efforts to learn as much as possible about the people and resources of the regions to be conquered, dominated, and perhaps settled, and to learn as much as possible about the regions already conquered in order to facilitate the administration and economic exploitation of these regions. The nineteenth century was the age of scientific exploration—Darwin in the *Beagle*, Livingstone in Africa, Powell in the Rockies, and so on—but the sources of support for these efforts tended to be institutions with very practical interest in the regions being studied. Paralleling all of this was the great surge of missionary activity that supported some exploration (including Livingstone's) but most crucially led to the gathering of important, detailed, information about ethnography, languages, and geography by hundreds of dedicated missionaries throughout the non-European world. Also of great importance were the detailed reports that colonial administrators everywhere were required to submit, reports providing information about native legal systems, land tenure rules, production, and much more.

Most of what was learned about non-Europeans came from these sources. It is not necessary for me to dwell on the fact that the people who supplied the information were Europeans with very definite points of view, cultural, political, and religious lenses that forced them to see "natives" in ways that were highly distorted. A missionary might have great love and respect for the people among whom he or she worked but could not be expected to believe that the culture and mind of these non-Christians was on a par with that of Christian Europeans. A colonial administrator not only had cultural distortions but usually worked with economic interests and classes (European planters, mining corporations, zamindars, and so on) and consciously or unconsciously put forward views about common people, and about resources, that reflected the biases and concrete interests of these elite groups. Strictly speaking, missionaries and colonial administrators were in the business of diffusing Europe to non-Europe. Thus the entire corpus of information about non-Europeans that was gathered in this process has certain quite definite distortions. It was immensely valuable in spite of this fact, but the plain fact is that theories constructed from this information—and this includes the great bulk of nineteenth-century anthropological, geographic, and politi-coeconomic theories about non-Europeans—are systematically distorted. The distortions are, broadly, those of diffusionism.

But colonial interest added an additional kind of distortion, a matter of shaping knowledge into theories that would prove useful for colonialism. Scientific and legal theories were constructed in general by policymakers and by intellectuals who were either themselves poli-cymakers or were close to policy. (In England, for instance, an extraordinary percentage of the influential historians, social theorists, even novelists and poets, had direct connections with the East India Company, the Colonial Office, and other private and public agencies of empire.[26]) I include under the label "theory" a wide range of general arguments, including the larger constructs of history.

At the most general level, intellectuals were shaping theories of social evolution which were in essence demonstrations that the postulates of diffusionism are natural law. As we noticed previously, the nineteenth-century debates between those called "evolutionists" and those called "diffusionists" were essentially debates between two versions of diffusionism. A great range of theories in both camps were constructs aimed to assist in colonial activity, and to develop and strengthen the doctrine that European colonialism is scientifically natural, a matter of the inevitable working out of social laws of human progress (development of the family, law, the state, etc.). Also at the very general level, seminal works of the middle and later nineteenth century on ancient and modern history presented various sophisticated diffusionist ideas, mainly about the reasons for and the facts of Europe's natural and persistent progress compared with other peoples, those now being colonized. These historical constructs were important in building support for colonial activities among European populations; and later, as colonial educational systems appeared, for convincing the natives that colonialism was natural, inevitable, and progressive.

In Chapter 2 I will discuss a number of these diffusionist theories which emerged, or became concrete, in the nineteenth century in the complex association of scholarship and colonialism, focusing on those theories that today underlie the myth of Europe's historical and cultural superiority. Here I need only show how classical diffusionism arose in tandem with classical colonialism, and for this purpose a pair of representative theories can be singled out by way of illustration.

One such theory was the postulate that non-Europeans have not developed concepts of private property in important material resources such as land. The theory asserted that private property emerged from ancient European roots, notably Roman land law and various putative Germanic traits relating to individualism; that other civilizations, lacking this history (and, by implication, lacking the mental and cultural qualities associated with this history), remained in a stage of evolution in which true individual ownership could not be fully conceptualized. These people, therefore, needed to have capitalism imposed on them. In fact, the theory was developed mainly by lawyers and administrators in the European colonial corporations and colonial offices, and had one very concrete purpose: to establish the legal basis for expropriating land from colonized peoples, on the fiction that the colonized had no property rights to this land because they had no *concept* of property rights in land.[27] Yet the theory became essentially an axiom in nineteenth-century intellectual thought. Even Karl Marx accepted it, and doing so produced a large theory about the evolution of private property—a major part of his theory about the origins of capitalism—in which it was argued (or rather assumed) that this evolution was peculiarly a European phenomenon, and that colonialism, for all its horrors, did at least bring about the diffusion of capitalism to the non-European world, a necessary though painful process for the non-Europeans. Thus even Marxism, perhaps the most antisystemic doctrine to come out of nineteenth-century Europe, was strongly shaped by diffusionism.[28]

A larger form of this doctrine, the more general "myth of emptiness," the diffusionist idea that a colonized or colonizable territory was empty of population, or was populated only by wandering nomads, people with no fixed abode and therefore no claim to territory, or lacked people with a concept of political sovereignty or economic property, had similar colonialist functions, and arose in a similar way. The same is true of a closely related doctrine, the theory of "Oriental despotism" (older in point of origin, but developed fully in nineteenth-century diffusionist thought), according to which non-Europeans lack the concept of freedom, hence suffer despotic governments that stifle all progress—until Europeans bring freedom to them in the form of colonialism (which ironically is the purest negation of freedom). These and other theories that arose from classical diffusionism are still employed today to reinforce the myth of Europe's historical and cultural superiority; we will discuss them in Chapter 2, which seeks to refute this myth.

The era of classical diffusionism was the era of classical colonialism, the era when European expansion was so swift and so profitable that European superiority seemed almost to be a law of nature. Diffusionism, in its essence, codified this apparent fact into a general theory about European historical, cultural, and psychological superiority, non-European inferiority, and the inevitability and absolute righteousness of the process by which Europe and its traits diffused to non-Europe. Diffusionism then ramified the general theory into innumerable empirical beliefs in all the human sciences, in philosophy, in the arts.[29] And it applied these beliefs in particular cases, explaining and justifying the individual acts of conquest, of repression, of exploitation. All of it was right, rational, and natural.

Modern Diffusionism

The nineteenth century, or more precisely the interval between the defeat of Napoleon and the beginning of World War I, was a time of relative peace and relative progress for Europeans. Colonialism fueled this process with resources, markets, cheap labor, and lands for settlement by Europeans, and colonialism resolved many of Europe's internal contradictions in the process. The idea that progress in European civilization and expansion of that civilization in space were different dimensions of the same historical force was the dominant idea of the time, and this was, of course, the central notion of diffusionism.

But all of this changed early in the twentieth century. The world is finite in size, so spatial expansion had to come to an end, and by 1900 all of the non-European world had been carved up into colonies, semi-colonial spheres of control, and territories of settlement. This change of conditions produced a change in thought: the essential problem now was exploitation and the maintenance of control in the face of native resistance. Thus it became a question not of expansion but of equilibrium. At the same time tensions among European powers—some of the tensions were connected to conflicts over colonies—boiled over into general war among the European powers. Soon after World War I came the Great Depression. Then, immediately, World War II. Between 1914 and 1945, then, the minds of European intellectuals were focused not on the idea of progress and expansion, but on the question of how to prevent disaster: how to maintain, or return to, peace and prosperity. The code word was "normalcy."

The central notion of diffusionism did not fit this intellectual mood. The prevailing doctrines of this period were theories of stasis, of equilibrium, not theories of expansion. Economics dwelled on Keynesian ideas of equilibrium. In geography, the doctrine known as "regionalism" prevailed, the idea that the various parts of the world are stable, coherent, well-demarcated regions and tend to remain that way. Anthropology was emphasizing two equilibrium theories: "functionalism," a model of social systems (and cultures) as stable and self-correcting systems, and "cultural relativism," a doctrine that declared in essence that each culture has intrinsic worth. Anthropologists of course worked primarily among colonized peoples, and these two theories were closely integrated with colonial policy, mainly as a basis for policies designed to prevent native unrest while allowing European exploitation of land, minerals, and labor.[30] Thus equilibrium doctrines were very widespread, probably dominant, in European thought throughout most of the first half of the twentieth century.

Diffusionism, in this period, seems to have gone into a partial eclipse. History and geography textbooks were still sublimely diffusionist, in an essentially nineteenth-century mood, emphasizing the beneficial spread of civilization to Africa, Asia, and Latin America ("where our bananas come from"), the teleological rise of "the West," and so on. In social thought, the doctrines of "extreme diffusionism" (discussed earlier) were still being advanced, and still debated.[31] And it should not be thought that the decline of diffusionism as a doctrine of cultural dynamics implied a decline in prejudice. The notion that non-Europeans are less rational, less innovative, and so on, was as intense as ever: perhaps even more intense since this was the period of Nazism and like doctrines, and since genetic racism seemed, in this era, to be science, not prejudice. We return to this matter in Chapter 2.

A new and modern form of diffusionism gained prominence after the end of World War II, in the period of collapsing colonial empires and an emerging "Third World" of underdeveloped but legally sovereign countries. This doctrine, generally known today by the title "modernization," or "the diffusion of modernization," arose in the late 1940s and the 1950s. Immediately after the Japanese surrender in 1945 it became clear that a number of colonies would gain independence immediately: liberation forces were now very strong and, in the wake of the world war, all of the colonial powers except the United States were now quite weak. All of them wanted to hold on to their colonies, the sources of great profit in the past and presumably also in the future. Each of the colonial powers maneuvered in its own way to hold on to its colonies, sometimes resorting to forcible efforts to suppress independence movements, sometimes conceding political independence grudgingly but peacefully where continued colonial control seemed patently impossible.[32]

All colonies had been saturated during the classical colonial era with the ideological message that economic and social progress for the colonial people had to come through the diffusion of "modernization" from the colonizing power. "Modernization" meant the diffusion of a modern economy (with major corporations owned by the colonizer), a modern public administration (the colonial political structure), a modern technical infrastructure (bridges, dams, and the like, built by the colonizer), and so on. I call this an ideological message, but it was in fact believed in profoundly by the colonizers, who felt that their mission was, indeed, to diffuse their own civilization to the peoples who were under their "colonial tutelage," and the fact that this mission produced wealth for their own country seemed only logical (recall diffusionist proposition 6). In the new situation the colonizers had to persuade the colonized that the "modernization" message was still valid. Doing so, they might convince the colonized to voluntarily relinquish the ideal of political independence in favor of the more pragmatic ideal of economic and social development under a wise and benevolent colonial rule. Or, if independence was insisted upon, this ideology would convince the people of the country now acquiring freedom that the only way to develop that country economically and socially was to retain the colonial economy, that is, to allow the colonizer's corporations and banks to continue their (profitable) work under the new regime: a system everyone today describes as "neocolonialism."

Now all colonial powers began a major campaign to intensify the process of colonial economic development.[33] This should not be thought of as cynical or hypocritical: remember that diffusionism defined the colonial process as beneficial for the colonized as well as the colonizer, and the technical and other personnel involved in the new colonial development activities were utterly convinced that they were working for the advancement of the colonized people. At the same time, a parallel campaign was developed to further the same form of economic development in the independent countries, partly through agencies of the United Nations, partly through bilateral aid agreements.[34] The United States, now the leading economic power, began to establish its own aid programs in countries throughout the underdeveloped world. Again, this should not be dismissed as cynical and political: there was, in this period, a tremen-dously euphoric ideology that saw the end of the world war as the beginning of an Age of Development, a time when the advanced nations would work to bring—that is, diffuse—prosperity and advancement to the poor nations.

Decolonization spread, and many liberation movements and newly independent countries refused to accept the neocolonial option, either ejecting foreign corporations (as did Indonesia) or opting for a specifically socialist society. This added a new impetus to the diffusion-of-moderniza-tion project. Efforts to bring about development through diffusion were intensified in hopes that their success would lead countries to reject the anticapitalist and antiforeign options. But the choice of either of these two options seemed to imply a political alignment with the Soviet Union and China, so the diffusion-of-modernization project now became important as a matter of foreign policy in the Cold War. In 1959 the Cuban revolutionary victory gave the project very high priority for the United States, which now treated modernization and economic development, particularly in Latin America, as a matter of the highest priority, calling for very large investment.[35]

Modern diffusionism is the body of ideas which underlay, and still underlies, this new set of conditions in the Third World. The diffusion of modernization, as it is carried out by public policy makers and private corporations and theorized by intellectuals (at least in the metropolitan, formerly colonizing, countries), is considered to be essentially the process by which Third World countries gain prosperity by accepting the continu-ous and increasing diffusion of economic and technological plums from the formerly colonial countries, a process that now, as in the past, is supremely profitable for the latter. In the diffusionist belief system, it is profitable for everybody, and also is right, rational, and natural, just as it had been a century earlier.

The ideas of 1993 are of course very different from the ideas of 1893, so it would be wrong to think that modern diffusionism is the same as classical diffusionism. Biological racism is no longer part of the model (as we will see in Chapter 2), and few modern European thinkers believe that non-Europeans simply do not have the potential to develop eventually to the level of Europeans. Religious undertones are largely absent, and the notion that a Christian god began things with the putative ancestors of Europeans, in the Bible Lands, and thereafter guided Europeans, Chris-tians, to persistent superiority over all others, is no longer very popular. The historical greatness of some non-European civilizations is now fully conceded. (But with one vital qualification: less rationality, less innova-tiveness, than European civilization. We deal with this in Chapter 2.)

After about a quarter-century of blind faith that the diffusion of modernization would bring about economic development everywhere, European
experts and scholars now qualify their belief in this model, and in
particular draw back from their former naive faith that the diffusion of
modern technology, particularly in agriculture, is the key to economic
development, the key to what used to be called "the take-off into sustained
growth."

But, all of this notwithstanding, the basic propositions of diffusionism
remain in place. Europeans still believe that Inside has one fundamental
cultural nature and Outside another, now admitting Japan into the Inside
sector. It is still believed that Europe in the past displayed a progressiveness
not found in any other civilizations, except in one or two places, at one or
two moments of history. Although European scholars no longer insist that
this fundamental difference between a progressive Inside and a stagnant or
slow-moving Outside will persist into the indefinite future, most of them
write and speak about the present and future as though this fundamental
dynamic will continue (again qualifying the picture to admit Japan, and
perhaps a few small East Asian societies, into the dynamic of Inside). But
today, as we will see in the next chapter, there is a growing, though still
small, group of European scholars, mainly responding to the newer ideas
that now emanate from non-European scholarship in postcolonial societies, who question the overall diffusionist model, and who deny its
historical conceptions about the superiority of Inside over Outside.

WORLD MODELS
AND WORLDLY INTERESTS

Diffusionism is a poor theory. Diffusionism, as I argue in this book, is not
good geography and not good history. Yet it exerts a tremendous influence
on scholarship, and has done so for a very long time. How do we account
for the fact that a bad theory can be so widely believed to be true, and for
such a long time? We should briefly consider this question before we shift
from the discussion of the nature and evolution of diffusionism as a theory
to the discussion of empirical history, the topic of later chapters. It is
important to understand how this (and every) theory interlocks with
other ideas and responds to social interests.

The Ethnography of Beliefs

In this discussion we will look at ideas as cultural facts: we will look at
them ethnographically, as beliefs held by human beings who belong to

specific sorts of communities and categories. We will see that a study of ideas-as-beliefs is quite different from an inquiry into the validity or truth of ideas, and that the study of ideas-as-beliefs is in some cases the more important and more basic of the two sorts of inquiry. We will see, in addition, that scientific beliefs arrange themselves in larger structures, belief systems, and that belief systems (like diffusionism) have certain crucial relationships of compatibility to one another and conformality to the values or interests of the groups of people who hold them to be valid. Pursuing this ethnographic exploration of the character of ideas-as-beliefs we will, I think, discover why the theory of diffusionism has had a life history that is more clearly explained by the life history of European society, and more particularly European colonialism, than it is by any intellectual or social process within the scientific community.

Scientific ideas, and empirical ideas in general, can be examined in two ways. One is comfortable and traditional. It considers ideas in terms of their communicated meaning. Are they logical; that is, do they reflect an internally consistent argument? Are they valid in the sense that what they assert about the real world seems to have evidential support? This combination of logic, or structure of argument, and evidential basis is the kind of thing we look for whenever we evaluate scientific ideas—indeed all ideas that concern empirical reality. The second way of looking at ideas inquires about the people who believe a given idea, who communicate it to others as a belief, and about the people who listen and in turn accept the idea as a belief. The question whether a person believes in the validity of an idea is not at all the same as the question whether the idea is in fact a valid one. Questions about belief status are matters of ethnography: of finding out why beliefs are held by given people; how beliefs come to be accepted and rejected by these people; how given beliefs are connected in the minds of these people with other beliefs held by them; how new candidate beliefs are weighed and accepted or rejected; and how beliefs as such are connected to other parts of culture, including values, social organization, class organization, politics, and so on. What makes this kind of inquiry threatening is the fact that it can provide independent and reliable evidence that a given group of people holds a given idea to be true for reasons that have little to do with logic and evidence, for reasons grounded in culture.

Interestingly, we experience no discomfort and sense no threat when we read an account written by some anthropologist or cultural geographer about the beliefs, values, myths, and so on, of some small and obscure society in some far corner of the earth. Indeed, we expect an anthropologist to tell us more about the social or cultural reason why the "natives" hold to these ideas than about the validity of the ideas. In this kind of context it is quite normal to have a description of ideas that distinguishes between the matter of their logical and evidential basis and the matter of their cultural binding. But when this ethnographic approach is applied to what are called "Western" ideas, in the realms of science, history, and the like, the results are disturbing and the enterprise itself seems somehow improper.

Ideas are, so to speak, surrounded by culture, and we can examine the surroundings and the way ideas are embedded in their surroundings. This is the ethnographic study of ideas. Now it happens to be a terminological convention in the field of anthropology to attach the prefix *ethno-* to a word designating a particular field of knowledge, such as medicine, botany, geography, and the like, when our purpose is to study that body of knowledge ethnographically. The study of "medicine," for instance, is different from the study of "ethnomedicine." The latter is an ethnographic field, asking what the medical beliefs are in a given culture, how these beliefs relate to the rest of that culture, and how to generalize cross-culturally about medical beliefs in all (or some set) of cultures. When we put together along with ethnomedicine all the other scientific fields prefixed by "ethno-," we get, naturally enough, *ethnoscience*, meaning the ethnographic study of all sciences; more broadly, all fields of empirical belief. Ethnohistory is part of that corpus.[36] So, too, is ethnogeography.

The subject matter of ethnoscience is *belief*. Ordinarily we look at beliefs as they are enshrined in empirical statements, generally sentences that assert some predicate to be true about some subject. The fundamental, though not the smallest, unit of study in ethnoscience is the statement of belief and the person or group who makes—and holds to—this belief statement. For every empirical statement in social science there is an ethnoscientific question about its *belief status* and there is a profoundly different question about its *truth status*. The two questions do not forever remain separate, but they come together only at the conclusion of a long analysis. That analysis explains in the end why so many diffusionist statements in which historians and geographers firmly believe are really false.

The study of beliefs is also the study of belief-holding groups. Two very important points have to be made about belief-holding groups. A belief-holding group can be a group of any type. However, among the various types of belief-holding groups, the most fundamentally important are cultures, classes, and combinations that can be thought of as ethnoclasses. None of these types is an abstraction, except in the matter of defining units and boundaries. Cultures are highly variable from place to place and person to person, but the analytical unit itself is real and concrete. Its individual members also have concrete reality. There is no philosophical conundrum about the cultural whole "versus" its individual (human) parts in the matters we are discussing now. Thus there is an ethnoscience of each individual human being and also an ethnoscience of groups as collectives. Classes are a bit more problematic. However, most people accept the broad idea that there is a basic division between two class communities, one, the working class, the other an elite class, deploying political power and generally accumulating wealth. I will simply take it as given here that a rough class bifurcation exists in most societies of this and the preceding century, while conceding that it is not always possible to tell whether a given group in a given society belongs to the working (producing) class or to the elite class or to some ambiguous or uncertain grouping that does not fit comfortably in either of the two. One such problematic grouping that relates closely to the issues discussed in this book includes professors and others engaged in studying and writing about society and the environment. They are not members of the accumulating class, the elite, but scholars and writers are in most cases (not all) strongly bound to that class, and for all their intellectual penetration, discipline, and honesty, they tend to think, say, and write down ideas that are useful to the elite. This is true most pointedly for ideas of the sort related to diffusionism.

Culture and class intersect in ethnoclass communities. There is a crucial use in this book for the concept of an ethnoclass community. I will argue strongly that the elite groupings of European countries together, in spite of their cultural (and national) differences, are a basic and permanent belief-holding group, and that their beliefs form, to a large extent, a single ethnogeography and ethnoscience. This reflects the fact that, for the period with which we are mainly concerned, the nineteenth and twentieth centuries, these elites have had a common set of interests in relation to the working classes of their own countries and of the non-European world, and they have together underwritten the production of a coherent belief system about the European world, the non-European world, and the interactions between the two. The most important proposition in this book, in fact, is the assertion that diffusionist ideas are at the core of the single belief system generated under the influence of, and for the interests of, the European elite. Although it should not be thought that the science and history produced under the stimulus of this ethnoclass is "biased," nevertheless we will see that the entire body of ideas concerning geography, history, and social science are strongly influenced by their ethnoclass patronage. Overall, and through the two-century-long flow of ideas, the product is a body of diffusionist beliefs which persist, and continue to influence social practice, although they are quite unscientific.

Beliefs assemble themselves into *belief systems*. The difference is not exactly a matter of relative complexity. Most simple beliefs are just those ideas that can be expressed in simple declarative sentences (although some call for a poem or a painting), and which people do express as assertions, with more or less confidence as to their veracity. We are interested in three things about these belief sentences and the acts in which they are expressed. They are empirical (not purely logical or purely evaluative). They are expressed as true, or possibly true. And they are thought of by the belief holders as cognitively whole or discriminable, as, putting it crudely, "concept," "ideas." What makes such a hard-to-be-precise-about unit important is the fact that individual humans do not simply think up such ideas or concepts out of thin air or immediate perceptual experience. These unit beliefs tend to retain their character as beliefs through long intervals of time, often generations, and among large numbers of people. Still more structure is to be found in a belief system that consists of chains of statements connected together by implication ("because . . ."). This kind of belief system I will call, unoriginally, an argument. It simply needs to be noted that the chain of statements in an argument may or may not proceed from simple to complex. A tighter structure is to be found in those systems called theories. Simple beliefs, then, are aggregated in various ways into belief structures, belief systems. Each belief system, in turn, is assembled (psychologically) into various sorts of higher order systems.

The highest level, comprehending all of the empirical beliefs held by a given belief-holding unit—including some beliefs that contradict one another—is the group's (or individual's) ethnoscience as a whole. Included here are all of the beliefs held to be true, or possibly true, about the external world, natural and social; beliefs about the self or person; and beliefs about technique—the self's capability to manipulate and influence the world. The ethnoscience, in a way, is an encyclopedia.

How does it happen that a new belief is admitted into a belief system? The question is not where new ideas come from but how they become *validated*, that is, given a kind of social license that admits them to the status of a belief, a belief that is at least accepted as a tenable hypothesis, a "reasonable idea," and at most accepted as fact. Three quite distinct judgments—I think also distinct procedures—are involved in this licensing process. One is a judgment of *compatibility*. The second is a judgment of *verifiability* (the matter of empirical verification). The third is a judgment of *conformality* (or value conformality). Scholarly conceits to the contrary notwithstanding, verification is the least important of the three.

All the belief systems held by a given group are in one way or another interrelated. Some are tightly connected: one theory may seem to follow directly from another, for example. Normally the relationship is much looser. But all belief systems have one basic and common relationship with all other belief systems held by a group. They are *compatible*. This means that they can coexist peacefully in the same ethnoscience: they are not cognitively or culturally dissonant. Although beliefs may occasionally contradict one another, total lack of relationship is not in principle possible: directly or indirectly, all beliefs held by a given belief-holding group (or individual) are somehow linked together, and some judgment of their compatibility will be made more or less often. Typically, belief systems reinforce one another: "If P is true, it is reasonable to suppose that Q is true." Or "If P is true, it follows necessarily that Q is true." What counts here, because this is an ethnographic scenario, is the fact that the belief systems are judged to be compatible. The judgment of compatibility is not simply a definition, an assertion to the effect that when the group holds two or more beliefs at the same time they are merely labeled "compatible." Compatibility is the outcome of an important social process. The process is most transparent when new candidate beliefs are introduced: when some new hypothesis is proposed within a belief-holding group and must, as it were, apply for a license. One of the most crucial tests it must pass is that of compatibility with existing beliefs.

Compatibility is the loosest of all relations among theories and other beliefs within an ethnoscience. Compatibility is, in a sense, a bridge that must be built over gaps in the overall body of thought. One sort of gap is obvious: it is the incompleteness of knowledge. The other sort, less obvious, is highly significant for the argument of this book. Here, the gap among theories (and other beliefs) is actually filled, but not by argument. In ordinary language, "it seems reasonable" to suppose that one theory lends support to another, or one historical belief is explained by another. This matter of "reasonableness" bears very close examination, because "reasonableness" is that form of the relation of compatibility which allows the most absurdly *un*reasonable ideas to pass for well-founded scientific argument.

Broadly speaking, there are two important ways in which gaps are bridged with "reasonableness" in place of—and disguising the lack of—an explicit, defensible argument. One way calls for an insertion of value statements in place of belief statements; we will explore this device in the next section. The other device lies within the belief system itself. It substitutes, for an explicit argument, an implicit one. To understand how it works we have to distinguish between *explicit* and *implicit* beliefs.

Implicit beliefs are usually matters "too obvious to mention," "beneath notice," "obviously true," "taken for granted." They are ordinary beliefs (and theories) which tend not to surface in discourse. There is nothing mysterious about most implicit beliefs; they are not somehow buried in the unconscious, or deliberately hidden from view. Some beliefs simply tend to be more often thought about consciously and verbalized than others, for a variety of unsurprising reasons. Those readily verbalized are explicit beliefs, those not, are implicit beliefs. There is no hard line of demarcation in a case of this sort, involving informal communication of belief.

But the demarcation is quite sharp in the case of formal, written and published, expositions of belief. The implicit beliefs are simply not written about. If they emerge in print at all, it is in the form of explicitly stated assumptions, or "axioms." Here, the conclusion that terminates an implicit argument is exposed to view, but not the argument itself. Rarely is there any attempt to deceive or obfuscate. The writer merely takes it for granted that the reader holds the same set of implicit beliefs and is willing to accept the "reasonableness" of unsupported assumptions. Both share the same ethnoscience and the same value system.

Implicit beliefs, as we will see, are the weakest link in the diffusionist world model. Throughout this book we will encounter diffusionist theories that simply hang in the air, unsupported by argument and evidence. In fact we will find, for large chunks of the diffusionist world model, that most of the propositions which would be needed to make these areas of belief coherent and sound are simply *missing*. In other words, relatively few of the beliefs are explicit and grounded. And these few are not connected together, so the fabric as a whole is incomplete. It seems complete only to those who are prepared to take a great deal for granted, who fill in the gaps with implicit beliefs.

Verification is the matter of testing a candidate belief to see whether it fits the facts. There are various kinds of test and various controversies about the nature of verification, but such questions need not detain us. The normal sort of verification involves a search for evidence that would seem to support or contradict the new hypothesis, the candidate belief. The process is never complete: everyone, of every culture and community, has to be satisfied with partial confirmation (and disconfirmation) of empirical beliefs. For us the important point about verification is this: verification is never sufficient grounds for converting a hypothesis into an accepted belief. Nor is it even necessary.

The judgment of compatibility is more crucial than that of verifiability, and this holds true among social scientists as it does among all other groups of natives. Part of verification is itself a matter of judging compatibility. The words and procedures used in verification, the criteria on which a test is deemed adequate, and much more besides, are drawn from the stock of existing beliefs, and the test of a candidate belief is therefore only partly a matter of direct confrontation with new evidence. But this is not the main point. In any belief-holding group, a new idea, a candidate belief, tends to be judged more on the basis of the way it fits into the existing belief system than on the basis of its directly apprehended meaning.

The same thing happens in self-conscious scholarship. It has long been a truism that existing scientific beliefs tend to be defended in the face of new hypotheses that question them, and the defense is often fierce, bitter, and dogmatic. (Whitehead once called scientists "the leading dogmatists. Advance in detail is admitted: fundamental novelty is barred."[37]) The truism was popularized by Thomas Kuhn in a dramatic scenario that he called a theory of scientific revolutions: in essence, crucial scientific beliefs gain a position of suzereignty, their supporters gain positions of academic power, and so the beliefs become entrenched and hold sway for long periods of time, even after evidence against these beliefs has massively accumulated. Eventually there is a historic break: a "scientific revolution" which overthrows the existing ruling theory or paradigm and installs another one, which, in turn, holds sway for a period of time.[38] But Kuhn's frame of reference (in his *Structure of Scientific Revolutions*) was physical science, and we are told very little about the historic process by which beliefs gain, and retain, hegemony in social science, including history and cultural geography. In these fields the process is fundamentally different. For one thing, influential beliefs hold sway for reasons that reflect much more directly the interests of elite groups outside of the scholarly field itself, and the replacement of one theory by another reflects mainly these external interests, not a Kuhn-like intellectual revolution within the community of scholars itself. Secondly, in these fields it is much less easy to attack old theories with new evidence, in part because scientific methods here are usually very inexact and in part because the collecting of evidence itself is guided and sometimes determined by existing beliefs. Thus it is that large belief structures, like diffusionism, persist for generations, untroubled by "scientific revolutions."

People have a tendency, sometimes slight, sometimes strong, to believe what they want to believe. Another way of putting this is to say that beliefs are influenced by *values*, or that cognition interacts with valuation to produce what Tolman elegantly called the "belief–value matrix."[39] The most straightforward (or at least modest) notion of "values" finds them to be judgments of preference, assertions of what is good and bad, right and wrong, liked and disliked, by individuals and groups. Values, like beliefs, are aggregated into systems. But value systems are very different from empirical belief systems. The latter, broadly put, assert things to be true (or untrue) about the world; the former assert things to be preferable (or not), hence they call for action upon the things uncovered by belief, and John Dewey was right in describing such assertions as "agendas."[40] Viewed in this way, the realm of value is not autonomous and opaque. It is mainly a transition zone between belief and practice. Values are *interests*.

Values interact with beliefs in a belief–value matrix. A belief system and a value system tend to maintain some degree of consistency with each other during limited periods (except in times of very rapid social change). This somewhat regular relation between the two systems I will call *conformality*, and it works both ways. Statements do not ordinarily become validated beliefs if they do not conform to the values, and therefore the interests, of the group. But value judgments indicate preferences for future action, and a given judgment is likely to be rejected by a group, sooner or later, if the future action it calls for is flagrantly impractical: if the action clearly cannot succeed given the nature of the real world as depicted in the belief system. Obviously, matters are more complex than this and also less predictable. In any case, the dominant belief system for a group must in the long run conform to the value system, and when the two fall out of conformality, one or the other will be forced to change. Since values are the expression of concrete worldly interests, the belief system will tend, in ordinary times, to bend more readily to the value system, and to worldly interests, than vice versa.

The judgment of value conformality is a crucial part of the binding of belief to culture. The notion that beliefs are culture-bound is of course a familiar one, but the idea that this proposition applies fully to the belief systems of scholars is not really accepted except in the most general and abstract way and therefore in a way that does not usually permit an analysis of the manner in which a particular scholar's values (or interests) affect his or her empirical statements. The stronger proposition, that all new ideas in social science are vetted for their conformality to values, and more precisely to the value system of the elite of the society—which is not necessarily the value system of the scholars themselves—and that this process of validation normally and frequently leads to the acceptance and persistence of really unscientific ideas, is hardly even considered, even though it is normal doctrine in ethnoscience (normal, at least, when applied to natives other than ourselves).

But there is an even stronger proposition that I will defend in this book with concrete arguments and evidence: the proposition that our world-scale models, and many of our specific theories and factual truisms, are accepted mainly—and in some cases only—because of their conformality to the values of the European elites; that this has been the case since the beginning of the nineteenth century, and is true today. Many of these demonstrably false beliefs are built into the world model of diffusionism, and this is so because diffusionism is the central intellectual doctrine that explains and rationalizes the actions and interests of European colonialism and neocolonialism.

Conformality is really the crucial part of validation. This is not ordinarily a matter of the "establishment" suppressing free speech (although this happens quite frequently). The judgment of conformality is a complex binding process. At all times the dominant group (a class or ethnoclass) has a fairly definite set of concrete worldly interests. Some of these conflict with others, but all tend toward the maintenance of the elite group's power and position. Because of its power to reward, punish, and control, this group succeeds in convincing most people, including most scholars, that its interests are the interests of everyone. These interests are social, economic, and political agendas, and it is a simple transformation to insert the word "ought" and turn them into values. Viewed statically, the interests are always clear, and the values derived from them cohere into a dominant value system that more or less mirrors these interests. Hence we have at all times a kind of environment of values, surrounding and influencing the ongoing validation process in scholarship.

The way this influence is exerted is very complex in our society today, but it was quite simple and transparent in the last century, when the main lineaments of diffusionism and other beliefs related to colonialism were being sketched in. In those days, not only was it true, as Marx said, that the ruling ideas are the ideas of the ruling class, but it was also true that almost nobody *but* the ruling class, and its subalterns, had the opportunity to render those ideas effective, in the form of publication, lecturing at influential schools and universities, and participating in policy formation and execution. Conformality, in those times and places, was accomplished largely through the social vetting process by which only those people who adhered to the dominant value system were in a good position to tender hypotheses as candidate beliefs. I do not believe that the process is altogether different today, but it would take us far afield to discuss it here at length. We can simply take note of the background of most professors (very few of whom are the offspring of poor or minority families), the reward structure in universities and consultantships, and other elements that jointly produce this result: few professional social scientists *want* to propose candidate beliefs which do not conform. This explains why, in spite of the most rigorous adherence to scientific method and scholarly canons, our theories remain, to a large degree, conformal.

In sum, validation proceeds by subjecting any candidate belief to three tests: compatibility, verifiability, and value conformality. Of these three, perhaps only conformality is essential (and then only in relatively tranquil times). Verification can be waived on occasion, provided that the hypothesis is both conformal to interests and nicely compatible with existing beliefs, be they explicit theories or implicit beliefs emerging as stated or unstated assumptions. Strong verification is, in any case, pretty hard to come by in the social sciences (a methodological fact that is often mistaken for an epistemological verity). In these fields, a judgment of compatibility is always rendered—if not on the candidate belief then on the person who proposes it—and it is most unusual for a new hypothesis or theory to become accepted as a belief if it contradicts the corpus of accepted beliefs in its field. But it is always the case that society and its elite need to be supplied with answers to pressing problems confronting them. So there is an important countercurrent. New hypotheses that display a touch of the novel and hold some possibility of solving an already recognized problem are encouraged, indeed rewarded. They must be compatible but not *completely* so.

There is simply no way that a scholar, once installed in the profession, can prevent conformal values from creeping into his or her work. This in spite of the fact that nearly all are honest, careful, and competent. The reason why my argument seems, incorrectly, to be an attack on scholarship is because the problem is almost always seen in connection with *explicit* beliefs, and with consciously held attitudes. The fields that study human society are a weak infusion of explicit theory within a body of belief that is largely implicit. Scientific method prevents us from accepting the arguments of an explicit theory merely because we want to do so. But the main grounding for each such theory is its compatibility with other beliefs in the system, expressed as a matter of "reasonableness." It is "reasonable" to accept certain assumptions (reflecting certain implicit beliefs) and not others. One theory seems "reasonable," or plausible, because it is compatible with another, accepted theory, although no explicit chain of connection exists: most links in the chain are buried in the realm of implicit belief. Finally, it seems "reasonable" to seek verification for a hypothesis with certain observations and not with others.

It should be added that the disciplined work of social scientists usually prevents them from unwittingly validating a hypothesis on grounds of value conformality. The problem, and it is a severe one, is that, thinking that our explicit beliefs are not validated by value conformality, we let value conformality control our implicit beliefs. These then provide a bridge across gaps between explicit theories, rendering them compatible, or serve up the assumptions which provide the starting point for new explicit theories, formal and informal. Scientific and scholarly method demands rigor only for the explicit theories.

Diffusionism as a Belief System

The discussion thus far has been designed to lay the foundation for an understanding of three aspects of the diffusionist belief system: its structure, its binding to certain groups in certain societies, and its evolution, this last aspect including the important questions of why diffusionism became prominent and why it has persisted. It would not be much of an oversimplification to say that diffusionism developed as the belief system appropriate to one powerful and permanent European interest: colonialism. From 1492 to the present the wealth drawn into Europe—meaning, as in all of this discussion, Greater Europe—from the non-European world has been a vital nutriment for the elite classes of Europe: for their maintenance of status within their societies, and for their progress. Whether this statement can be generalized to include all classes within European society, whether, that is, colonialism was an interest of the nonelite classes of Europe in most times and places, is a contentious issue which I do not need to address. I am merely asserting that (1) Europe's elite depended on colonialism; (2) Europe's elite was tremendously influential in the evolution of European ideas and more specifically European scholarship; and (3) Europe's elite held a permanent social interest in the creation and development of a conformal belief system, a body of thought that would rationalize, justify, and, most importantly, assist the colonial enterprise. As that enterprise evolved and changed, so too did the body of ideas constituting diffusionism.

Most of this book is devoted to a delineation and critique of diffusionist beliefs, so I do not need to review the nature of the belief system here. Suffice it to say, the doctrine ranges over all scales of fact, from world geography and world history to ideas about the qualities of individual human beings, European and non-European, and descriptions and explanations of particular local events. The scope of the diffusionist belief structure encompasses a fair share of European ethnoscience. That is, a fair share of the licensed belief statements in European ethnoscience are used within the diffusionist belief structure, although they are also used in other structures. (Some examples to be discussed in Chapter 2 are: beliefs about demographic behavior, about intelligence, about the origins of civilization, about the fertility of tropical soils.) These statements range from modest assertions of evidential fact to complex and elaborate theories, both formal and informal. They enter the diffusionist canon through all of the licensing procedures for belief acceptance discussed above. Over time, all of them go through the screening process of conformality with the value or interest of colonialism, or go through the indirect screening process that accords them the status of being compatible with other beliefs which are themselves conformal. Over time, the belief system accretes new diffusionist beliefs and discards those that contradict the canon or that have lost their relevance in a changing world. Since colonialism, in various newer forms such as neocolonialism, remains an interest of the elite, and since implicit beliefs go unnoticed and uncriticized, the process of adding, subtracting, and modifying diffusionist beliefs continues in the present. Were this not the case, the multitude of beliefs about Eurocentric history which claim our attention in this book would long since have been discarded.

Diffusionist beliefs at the space–time scale that embraces the whole world and the whole of history, tend to form a rather tightly structured theory, as we saw earlier in this chapter. The theory, in brief, describes the essential processes that take place in an "inner," essentially European, core sector of the world, describes those that take place in an "outer," essentially non-European sector, and describes the modes of interaction between the two sectors, the most important of which is the inner-to-outer diffusion of innovative ideas, people, and commodities.

We can describe this world-scale space–time theory as the "diffusionist world model." It is the colonizer's model of the world.

The obvious question arises: what would we conceive to be a *non*diffusionist world model? This would be a world in which the processes at work in any one sector are expected also to be at work in the other sectors. In essence, this model is driven by a concept of equal capability of human beings—psychological unity—in all cultures and regions, and from this argument it demands that any spatial inequalities in matters relating to cultural evolution, and more specifically economic development, be *explained*. Stated differently: equality is the normal condition and inequalities need to be explained. Diffusionism, in contrast, expects basic inequality between the Inner and the Outer sectors of the world—and of humanity. The uniformitarian principle is not one of uniformity; it is the principle of human equality.

At space–time scales smaller than the world, the diffusionist belief system is very diffuse, parts of it hanging together as formally elegant theories, parts of it floating around as compatible but weakly connected belief statements. It would take us far beyond the scope of this book to attempt a description of *all* of the parts, levels, and subsystems in the diffusionist model as a whole, ranging upward and downward from the world model to the level of particular space–time event descriptions. But we can make a start.

NOTES

1. In this book the word "Europe" refers to the continent of Europe and to regions dominated by European culture elsewhere, regions like the United States and Canada.

2. This quick survey of 150 years or so of world history textbooks is, of course, very schematic and impressionistic. Some further comments may be of use. In textbooks of the first period, around 1850 (plus or minus a decade or so), the original home of Man is often stated to be the Garden of Eden, which is located by different textbook writers in different parts of western Asia. For instance: somewhere east of Canaan and near Mesopotamia (Robbins, *The World Displayed in its History and Geography*, 1832, p. 13); somewhere in the "healthful" mountains between the Caspian Sea and Kashmir or Tibet (Müller, *The History of the World to 1783*, 1842, pp. 27, 43–44); perhaps near the borders of the Mediterranean Sea (Tytler, *Universal history, From the Creation to the beginning of the Eighteenth Century*,vol. 1, 1844, p. 17); in the Vale of Kashmir (Willard, *Universal History in Perspective*, 1845, p. 34); somewhere between the Caucasus Mountains and the Himalayas (Keightley, *Outlines of History*, 1849); in the Himalayas (Weber, *Outlines of Universal History*, 1853, p. 6); in Armenia (Collier, *Outlines of General History*, 1868). There seems to be a kind of median location of Eden near the Caucasus Mountains, which was also, by no coincidence, the supposed place of origin of the "Caucasian race." Noah, of course, began postdeluvian history on Mt. Ararat, in Armenia (also roughly in the region of the Caucasus). Noah is supposed to have migrated then to Europe (Whelpley, *A Compound of History, From the Earliest Times*, vol. 1, 1844, p. 10), or Mesopotamia (Robbins, 1832, p. 20), or Palestine, or some other part of the Bible Lands. Noah's three sons are supposed, then, to have dispersed and to have founded the branches of mankind—the first great diffusion process. In most textbooks of this period history tends to move west; some textbooks present the Hegelian notion that history proceeds inexorably westward, following the sun, with the implication that the United States, farther west still, will replace Europe as the next center of world civilization.

It was widely believed in this period that nonwhites were not truly and fully human. One version of this theory, the notion of "polygenesis," claimed that God had created true humans in the Garden of Eden and other races—or at least the "black race"—in other places and times. This theory questioned the standard interpretation of the Old Testament (that everyone is descended from Adam and Eve), so, not surprisingly, it was not stated as truth in the textbooks I consulted. (I have not, however, looked at textbooks used in the antebellum South, and the theory of polygenesis was most popular in slaveholding regions, as an ideological grounding for the treatment of blacks as things rather than people.) Yet polygenesis was important enough to be mentioned and then rejected—in favor of the view that all humans are descendants of Adam and Eve— in some textbooks, down to the end of the century (see, for example, Dew, *A Digest of the Laws, Customs, Manners, and Institutions of the Ancient and Modern Nations*, 1853; Fisher, *Outlines of Universal History*, 1885; Duruy and Grosvenor, *A General History of the World*, 1901). But polygenesis was not needed; the belief that nonwhites are inferior to whites is asserted in one way or another in all the textbooks I examined. The theory of "degeneration" served just as well as polygenesis. It was the notion that the descendants of Ham, and perhaps other biblical peoples, migrated away from the Bible Lands, eastward and southward, and degenerated from civilization toward savagery, or even lower, as they did so, because they had not accepted Christ, or because they migrated into inferior environments, or for some other reason. (See, for instance, Keightley, 1849, 5–6: "the savage is a degeneration from the civilized life," and Africans are near the apes.) Degeneration was asserted in some textbooks, but in most the simple fact of white superiority was stated and left unexplained. The history of the world is, in general, the history of the white race, or the Semitic and Aryan peoples (see below). For times later than the Roman era, non-Europe is scarcely discussed, except as a backdrop for discussions of the Crusades, the building of colonial empires, and the like. (See Harris, *The Rise of Anthropological Theory*, 1968, for an excellent discussion of polygenesis and degeneration.)

Nearly all textbooks of the period around 1900 (give or take a few years) accepted the newer scientific theories about the age of the earth and the fact of biological evolution (though not the Darwinian theory of evolution). The biblical account of human history, however, was retained in many books, although fewer of them accepted the formerly standard Old Testament chronologies (for instance, that things began in 4004 B.C.). Books of this period tended to present the so-called "Aryan theory," a theory derived from philology but expanded into a theory of culture history. Earlier philologists had identified an "Aryan" or "Indo-European" language family, and also a "Semitic" family (which many authorities, including textbook writers, identified with Noah's sons Japheth and Shem, respectively). The white race consists of these two peoples. One branch of the Aryans supposedly migrated west into Europe (from the supposed "Aryan hearth," somewhere southwest or northwest of the Caucasus). These were progressive, energetic people, who founded European civilization mainly after they had acquired Christianity from the Semites, who invented the first barbaric civilizations and monotheism but then stagnated into a dreamy, decadent, unambitious culture and thereafter ceased to advance civilization. No other culture, apart from these two, had much to do with history. (According to Freeman, *General Sketch of History*, 1872, p. 2, history "in the highest and truest sense is the history of the Aryan nations of Europe"; see also Collier, 1868; Swinton, *Outlines of the World's History*, 1874; Gilman, *First Steps in General History*, 1874; Anderson, *New Manual of General History*, 1882; Steele and Steele, *A Brief History of Ancient, Medieval, and Modern Peoples*, 1883; Fisher, 1896; Quackenbos, *Illustrated School History of the World*, 1889; Thalheimer, *Outline of General History for the Use of Schools*, 1883; Sanderson, *History of the World from the Earliest Time to the Year 1898*; Duruy and Grosvenor, 1901. Ploetz and Tillinghast presented this theory in *Epitome of Ancient, Medieval, and Modern History*, first published in 1883, in its many editions down to 1925 when, in an edition edited by H. E. Barnes, the theory was finally eliminated.) See Bernal's *Black Athena* (1987, 1991) for an insightful discussion of the Aryan theory and related topics in the history of European thought.

Because the Old Testament spoke of agriculture—Cain knew farming and

Abraham herded domesticated animals—the history textbooks tended not to address the problem of where agriculture was invented until very late in the nineteenth century, the period when science was beginning to deal with this problem. Some scientists and some textbook writers then began to speculate that, just possibly, agriculture was as old in continental Europe as it was in western Asia and Egypt. (From the point of view of science, see Joly, *Man Before Metals*, 1897.) But the idea that agriculture originated in the Bible Lands remained dominant although now it was viewed (by most) as an invention, not an artifact of original creation. The ethnographic fact that some tribal peoples (for instance, in Australia) did not practice agriculture was commonly explained, in the textbooks of the earlier nineteenth century, in terms of the theory of degeneration: their ancestors had somehow lost the art. Later in the century it became more common to use the diffusionist conception that agriculture had been invented by west Asian or (conceivably) European peoples, then diffused outward over the rest of the world, and cultures that did not practice it in modern times simply had not yet acquired it, either because of their isolation or because they were too stupid to take it up.

The Orient Express was a famous train that ran between western Europe and western Asia. Although various routes were used in different eras, the basic line ran from Constantinople (Istanbul) through Greece to northern Italy or Austria, then on to France, and (via Ostend) England. Most of the history textbooks write about world history as though it marched northwestward, rather like the westbound Orient Express, with stations (so to speak) in Athens, Rome, Paris, and London. (See Chapter 2 for further discussion of the Orient Express model.)

3. In nineteenth-century world history textbooks, Turkey was given some attention because of its political involvement with European affairs. World geography, in contrast to world history, always covered the entire world, and textbooks as well as the great multivolume descriptive geographies (like Reclus's classic 19-volume *Nouvelle Géographie Universelle*, published between 1876 and 1894) gave considerable attention to Asia, Africa, and Latin America. This should not mislead us, however. One of the primary functions of geography throughout this entire period was to teach European children what they needed to know about non-Europe in order to participate in their countries' imperial and commercial activities in these regions. See Hudson, "The New Geography and the New Imperialism: 1870–1918" (1977) and McKay, "Colonialism in the French Geographical Movement" (1943), on the close relation between geography and colonial activities.

4. The character of this newer approach can be seen if we look at two well-known modern university-level texts, one written by W. H. McNeill of the University of Chicago (*A World History*, 3rd ed., 1979), the other by J. M. Roberts of Oxford University (*The Hutchinson History of the World*, 2d ed., 1987; published in the United States as *The Penguin History of the World*). For pre-Christian-era world history, more than three-quarters of the place-name mentions in both books are places in Europe and the Middle East (including North Africa); less than one-quarter of the place-name mentions are places in other parts of the world and only about 1% are in Africa. For the Christian era to A.D. 1491 there is significant divergence between the two books. In Roberts's text, European and Middle Eastern places constitute about 85% of place-name mentions. In McNeill's text, European and Middle Eastern places constitute only 60% of the place names mentioned, a significant departure from the older tradition although still well out of proportion to this region's size in area and population. (Sub-Saharan Africa accounts for only 2% of mentions for the period

A.D. 1–1491.) In both books, therefore, the region I have been calling "Greater Europe" has considerably lower salience for pre-1492 history than was typically the case in older world history textbooks. As to *explanation*, however, both books retain much of the traditional perspective. Roberts gives almost no causal role to the cultures and regions of the world other than Europe and the Middle East (including North Africa) for any period prior to 1492. McNeill gives considerable weight to East Asia and some to South Asia for certain historical periods, but almost all of the world history-making forces emanate from Europe, western Asia, and North Africa for the period before 1492. (An exception is the Black Death, which, according to McNeill, swept westward into these regions from farther Asia. On this matter, see also McNeill's *Plagues and Peoples*, 1976.) Examples of the Eurocentric explanations offered by both authors will be given in Chapter 2 of the present book.

It is, however, insufficient to look only at those present-day textbooks that are clearly identified as "world history" textbooks. Quite frequently university courses covering the subject of world history use history textbooks that carry titles like "The History of Western Civilization." (See, for instance, Lerner, et al., *Western Civilizations: Their History and Their Culture*, 1988; Kagan, et al., *The Western Heritage*, 1987; Chambers, et al., *The Western Experience*, 1987.) If such a textbook neglects the non-Western world, there can be no complaint that it is misleading: the title clearly specifies "West" not "World." But if the *course* is designated as "world history" and the *textbook* is designated as "Western history," then we have a problem. The worst scenario would be one in which world-history teaching continues to be Eurocentric history in disguise. I do not know of any research which tests the following hypotheses: (1) Given that, today, historians are sensitive to the need to avoid Eurocentric bias in the teaching of world history, do some of them simply change the title to "Western" so that Eurocentrism becomes, then, licit? And (2) is it possible that there is a trend away from the teaching of "World" history and toward "Western" (etc.) history and that this trend reflects a reaction (or adjustment) to present-day demands for nonethnocentrism and "fairness?"

5. A school textbook is truly a key social document, a kind of modern stele. In the typical case, a book becomes accepted as a high school (or lower-level) textbook only after it has been reviewed very carefully by the publisher, school boards, and administrators, all of whom are intensely sensitive to the need to print acceptable doctrine; they are concerned to make it certain that children will read only those facts in their textbook which are considered to be acceptable *as* facts by the opinion-forming elite of the culture. The resulting textbook is, therefore, less an ordinary authored book than a vetted social statement of what is considered valid and acceptable for entry into the mind of the child. For this reason, research on textbooks (including college textbooks, in which the same process is at work, though more subtly) is, in fact, ethnographic research. It tells us about the belief system of the opinion-forming elite of the culture as a whole. Therefore, geography texts in the United States are really ethnogeography documents. Likewise, history texts are really ethnohistory documents. They are probably as useful as cultural artifacts as any old potsherd or inscription. See the final section of this chapter.

6. This argument is given much weight by Eurocentric Marxists, since class struggle, for Marxists, is the central force in historical evolution.

7. A form of political feudalism is sometimes conceded to have been developed earlier by the Chinese, but the great majority of European scholars, including most Marxists, believe that European feudalism was unique in representing a form of society that was a crucial, essential, stepping stone to modernity. See Chapter 3.

8. See Samir Amin's *Eurocentrism* (1988) for an excellent discussion of this notion. The word "Eurocentrism" apparently was coined quite recently, to assemble "European ethnocentrism" into one word. However, I (like Amin) do not think of Eurocentrism as merely a species of ethnocentrism, as the following paragraphs will make clear.

9. I use the word "community" to refer to any social unit, of any size. In this discussion, for the sake of simplicity, the "communities" will be thought of as villages distributed across a rural landscape. I neglect here the cases in which culture change results from the combination of an independent invention and a diffusion event. See my "Two Views of Diffusion" (1977) and "Diffusionism: A Uniformitarian Critique" (1987a).

10. See Jett, "Further Information on the Geography of the Blowgun and Its Implications for Transoceanic Contact" (1991); also see Carter, *Man and the Land* (1968), and Edmonson, "Neolithic Diffusion Rates" (1961).

11. See, for instance, Eliot Smith, *The Diffusion of Culture* (1933), Perry, *The Primordial Ocean* (1935), and Taylor, *Environment and Nation* (1945). Eliot Smith flatly asserts that the ancient diffusion process radiating mainly from Egypt and Phoenicia "continued for many centuries to play upon the Pacific littoral of America, where it is responsible for . . . the remarkable Pre-Columbian civilizations" (quoted in Zwernemann, *Culture History and African Anthropology*, 1983, p. 15).

12. On these matters, see Harris, *The Rise of Anthropological Theory* (1968) and Steward, *Theory of Culture Change: The Methodology of Multilinear Evolution* (1955).

13. See Koepping, *Adolf Bastian and the Psychic Unity of Mankind* (1983), Stocking *Race, Culture, and Evolution* (1968), and Harris, *The Rise of Anthropological Theory* (1968).

14. Often the antidiffusionist camp was labeled the "cultural evolutionist" camp, and the debate as a whole was labeled "diffusionism versus evolutionism." But, as I argue here, evolutionists were to some extent diffusionists and diffusionists were to some extent evolutionists. Moreover, I want to use the term "cultural evolution" in this book in a much broader and much less controversial sense, as indicating merely the search for explanation in larger questions of historical and cultural change. A problem that is "historical" becomes a problem of "cultural evolution" when we ask, broadly, "why?" Some scholars, of course, are not comfortable with this usage of the phrase "cultural evolution." For some it carries baggage of economic determinism or environmental determinism or technological determinism, or it signifies the notion of an invariant sequence of cultural stages through which all human groups must pass—but I mean none of that. Most cultural geographers use the phrase "cultural evolution" just about as I use it here.

15. Both forms are sometimes combined; for instance, in the nineteenth century northwestern Europe was considered (by northwestern Europeans) to be absolutely civilized, Africa absolutely uncivilized, and all other areas (of the Eastern Hemisphere) somewhere in between. These matters are discussed later in this book.

16. Max Weber's notion of "European rationality" is discussed in Chapter 2.

17. The varying conceptions of China, and also India, as semirational or intermittently rational or rational in some ways but not in others are discussed in Chapters 2 and 3.

18. The "bogeyman" refers to the Buginese, a Malay people who fought fiercely against the Europeans and so were stigmatized in this way. The most famous fictional vampire, Count Dracula, came to England from Outside (a barbarous mountain region on the frontier of the Turkish empire).

19. On the history of the idea of progress in European thought, see, for instance, G. H. Mead, *Movements of Thought in the Nineteenth Century* (1936), Toulmin and Goodfield, *The Discovery of Time* (1965), Nisbet, *History of the Idea of Progress* (1980), and Bowler, *The Invention of Progress* (1989). It is true that the idea of progress as the normal condition was doubted by some thinkers during the nineteenth century (and especially in opposition to the idea of biological evolution), but this was a minor and intermittent countercurrent. See Stocking, *Race, Culture, and Evolution* (1987) and Bowler, *The Invention of Progress* (1989).

20. See Huddleston, *Origins of the American Indians: European Concepts, 1492–1729* (1967), Williams, *The American Indian in Western Legal Thought* (1990); Hulme, *Colonial Encounters: Europe and the Native Caribbean 1492–1797* (1992); Gossett, *Race: The History of an Idea in America* (1963).

21. At times the notion was advanced that these civilizations had somehow evolved in antediluvian times and had not been wiped out in the Flood (see, for example, Keightley, *Outlines of History,* 1849). Haskel (*Chronology and Universal History*, 1848, p. 9) speculates that Noah migrated to China where he or his descendants founded the Chinese monarchy. "The early improvement and populousness of the east, seems to favor this idea."

22. From the mid-nineteenth century or perhaps earlier, the literature on the occult, on ghosts and monsters, and the like, tended to focus on extra-European origins or homes or sources of the witches, monsters, demons, zombies, walking mummies, evil spells ("black magic"), artifacts with supernatural powers, and so on, all of which have a tendency to diffuse into Europe as a kind of counterdiffusion, an undertow beneath European expansionism. See Brantlinger, *Rule of Darkness: British Literature and Imperialism* (1988).

23. See W. A. Lewis, ed., *Tropical development, 1880–1913* (1970).

24. See Bowler, *The Invention of Progress* (1989), Stocking, *Victorian Anthropology* (1987), Mandelbaum, *History, Man and Reason* (1971).

25. Spencer, *The Man Versus the State* (1969). The complex interplay between individualistic and holistic theories of historical progress during the nineteenth century is discussed in G. H. Mead, *Movements of Thought in the Nineteenth Century* (1936) and Mandelbaum, *History, Man and Reason* (1971). I try to show in Blaut, *The National Question: Decolonizing the Theory of Nationalism* (1987b) how most theories of nationality and national evolution emerge from one or the other of these intellectual streams, one essentially Kantian and psychologistic, the other essentially Romantic and Hegelian.

26. Among them: Malthus, J. S. Mill, T. Macauley, and Thackeray. See Brantlinger, *Rule of Darkness: British Literature and Imperialism, 1830–1914* (1987). Also see Williams, *British Historians and the West Indies* (1966) and Said, *Orientalism* (1979).

27. See Thapar, *Ancient Indian Social History: Some Interpretations* (1978) and "Ideology and the Interpretation of Early Indian History" (1982), and B. Chandra, "Karl Marx, His Theories of Asian Societies, and Colonial Rule" (1981). We return to this issue in Chapter 2.

28. See in particular Marx's article, "The British Rule in India" (1979). In later work, Marx and Engels adopted a much more negative opinion about colonialism, developing to some extent the idea of colonial underdevelopment: see Blaut, *The National Question: Decolonizing the Theory of Nationalism* (1987b) for a discussion.

29. See Chapter 2.

30. See Asad, *Anthropology and the Colonial Encounter* (1975), and Temu and Swai, *Historians and Africanist History* (1981).

31. Important examples are Eliot Smith *The Diffusion of Culture* (1933), Perry, *The Primordial Ocean* (1935), Schmidt, *The Culture Historical Method of Ethnology* (1939), Griffith Taylor, *Environment and Nation* (1945). See critiques in Radin, *The Method and Theory of Ethnology* (1965), Lowie, *The History of Ethnological Theory* (1937), and Harris *The Rise of Anthropological Theory* (1968).

32. Here I am of course rejecting the view, common among European scholars, that colonies were relinquished voluntarily. This view has been rejected almost universally by scholars in the formerly colonial world. It may conceivably have been true for a few small islands from which profits no longer flowed to the colonizing power, but even such cases are debatable. It is worth noting that the United States has not given independence to any of the colonies it held at the close of World War II; has not even formally conceded the right of full self-determination (including independence) to Puerto Rico, the Virgin Islands, the Marianas, etc. The other colonial powers would presumably have taken the same position, had they had the power to do so in the face of colonial pro-independence forces; in such cases as the Dutch East Indies, French Indochina, Kenya, Angola, Mozambique, and so on, the colonizers tried to hold on to their possession by force of arms and failed. My view is set forth in Blaut, *The National Question* (1987b), chap. 4, and Blaut and Figueroa, *Aspectos de la cuestion nacional en Puerto Rico* (1988).

33. Representative examples include "Operation Bootstrap" in Puerto Rico, the Colonial Development and Welfare programs in parts of the British Empire, the increased funding for colonial agriculture and health departments, the establishment of colonial universities, and so on. These programs, regardless of their underlying political purposes (often hidden from the technical personnel involved) were, overall, very impressive.

34. Often this work was a direct continuation of colonial technical work, often with the same personnel, now "foreign advisor" or "United Nations expert" rather than "colonial technical officer."

35. Thus the Alliance for Progress, the Peace Corps, the elevated funding of technical and financial agencies of the Organization of American States, and the like.

36. On ethnoscience, see, for example, Conklin, "Lexicographical Treatment of Folk Taxonomies" (1969), Frake, "The Ethnographic Study of Cognitive Systems" (1969), Blaut, "Some Principles of Ethnogeography" (1978), and Spradley and McCurdy, *Anthropology: A Cultural Perspective* (1975). In my view, the categories "history" and "science" cannot be distinguished ontologically, although historiography is hardly an exact science.

37. Whitehead, *Science and Philosophy* (1948), p. 129.

38. Kuhn, *The Structure of Scientific Revolutions* (1970). More relevant is Fleck's *Genesis and Development of a Scientific Fact* (1979).

39. Tolman, "A Psychological Model" (1951).

40. Dewey, "The Logic of Judgements of Practice," in his *Essays in Experimental Logic* (1916).

CHAPTER 2

The Myth of the European Miracle

M ost European historians believe in some form of the theory of "the European miracle." This is the argument that Europe forged ahead of all other civilizations far back in history—in prehistoric or ancient or medieval times—and that this internally generated historical superiority or priority explains world history and geography after 1492: the modernization of Europe, the rise of capitalism, the conquest of the world. Most historians do not see anything miraculous in this process, but the phrase "the European miracle" became in the 1980s a very popular label for the whole family of theories about the supposedly unique rise of Europe before 1492. The phrase acquired its new popularity mainly from a book by Eric L. Jones which appeared in 1981, a book simply entitled *The European Miracle*.[1]

The historians do not agree among themselves on the question *why* the miracle occurred: why Europe forged ahead in this perhaps miraculous way. Is it because Europeans are genetically superior? are culturally superior? live in a superior environment? Is it because one special, wonderful thing happened in Europe, or happened to Europeans at a special moment in history, giving Europeans a decisive advantage over other societies?

Nor do the historians agree about *when* the miracle occurred or began. Did it occur back in the prehistoric age in what some still call the "Aryan" or "Indo-European" culture? in the late-prehistoric "European 'Iron Age'?" Did it begin with the Greeks? with the Romans? in the early Middle Ages? in the late Middle Ages? Did it happen continuously throughout history—a series of miracles—each pushing the Europeans farther and farther ahead of the rest of humanity?

The historians debate these matters, the questions "why" and "when," but not the question "*whether*"—whether a miracle happened at all. Or, to be more precise, they do not even consider the possibility that *the rise of Europe above other civilizations did not begin until 1492*, that it resulted not from any European superiority of mind, culture, or environment, but rather from the riches and spoils obtained in the conquest and colonial exploitation of America and, later, Africa and Asia. This possibility is not debated at all, nor is it even discussed, although a very few historians (notably Janet Abu-Lughod, Samir Amin, Andre Gunder Frank, and Immanuel Wallerstein) have come close to doing so in very recent years.[2]

My task in this chapter and in Chapter 3 ("Before 1492") is to show that Europeans indeed had *no* superiority over non-Europeans at any time prior to 1492: they were not more advanced, not more modern, not more progressive. Then in Chapter 4 ("After 1492") I will show how colonial riches brought about the rise of Europe and led to Europe's ultimate hegemony over the world, showing also that Europe's internal characteristics do not explain 1492—do not, that is, explain the origins of colonialism.

There seem to be two basic ways to argue that the myth of the European miracle is wrong, that Europe did not surpass other world civilizations before 1492. The best way by far is to look at the facts of history, and demonstrate that the evolutionary processes that were going on in Europe during and before the Middle Ages were essentially like the processes taking place elsewhere in the world in terms of rate and direction of development. I will try to demonstrate precisely this in Chapter 3, which compares the medieval landscapes of Europe, Africa, and Asia, and shows how a transition from feudalism and toward capitalism was occurring in many parts of the Eastern Hemisphere just prior to 1492.

But the myth of Europe's unique "rise," its "miracle," is so deeply embedded in European historical thought that an ordinary argument from the facts probably would not be persuasive. As we saw in Chapter 1, the dominant theory has been defended by generations of historians, with practically no dissenting argument; it is supported as well by many other ideas which are accepted as unquestioned truth in European culture; and it fits in with and supports the interests of European countries (and corporations) in their dealings with the non-European world. For these reasons, I have decided to use another kind of argument—to demonstrate the fallacies in the dominant theory—as a sort of ground-laying for the empirical argument.

In the present chapter I will examine the most common arguments used by historians today to *support* the theory of the European miracle, and will try to show that they are unconvincing. This task is rendered somewhat complicated, for a small book like the present one, by the sheer number of different arguments currently in circulation and the number of historians who are writing books and articles on the subject of, and in support of, the theory of the European miracle. How, then, to proceed? I will advance by stages. First I will present a brief discussion of the ways in which historians have tended to argue the myth of the European miracle in recent decades, and I will show how a critical, revisionist point of view has begun to appear. Next I will lay out, in a kind of menu or classification or checklist, the most important arguments in support of this myth that are being put forward today, and I will try to show, for each argument in turn, how unconvincing it really is. In the third stage, I will summarize the empirical argument against the "miracle" position in two parts (the topics of Chapters 3 and 4, respectively): the evidence that Europe was not ahead of Africa and Asia (at a continental scale of attention) prior to 1492, and the evidence that colonialism after 1492 accounts for the selective rise of Europe.

MYTHMAKERS AND CRITICS

The idea that Europe was more advanced and more progressive than all other civilizations prior to 1492 was the central idea of classical Eurocentric diffusionism, as we saw in Chapter 1. Therefore we do not have to consider the origins of the European miracle theory: it is our inheritance from earlier times. However, after World War II the doctrine assumed a distinctly modern form. First of all, the racist arguments had been decisively rejected: no longer was it argued that non-Europeans are genetically inferior to Europeans and that it is this inferiority that explains why they lagged behind in history. Historians now generally accepted the idea that European historical advantages reflected facts and happenings of much earlier times. European superiority was a matter of prior arrival by Europeans at a stage of development that all other people could aspire to reach in the future: a matter, therefore, of priority, not innate superiority. Second, there was a rapid increase in historical scholarship concerning the non-European world after 1945, grounded in rather pragmatic political and economic interests, and emanating in part from government- and foundation-sponsored "foreign area studies programs," but this scholarship nonetheless added greatly to the knowledge available in Western countries concerning non-Western history. The new knowledge was fairly quickly put to rather limited use: some of the wilder fables were discarded but the basic ideas about the non-Western world did not significantly change. Third, and most importantly, the postwar world came to embrace the crucial new theory of "modernization," the theory that the diffusion of European ideas, things, and influence would bring about the economic development of the non-Western world in the coming Age of Development. This theory had important effects on history writing.

Modernization as History

The theory of modernization addressed the present and the future but it was fundamentally historical. Its basic principle was the notion that whatever had led in the past to European superiority could now be diffused out into the non-European world and assist that world to more or less catch up. As we saw in the last chapter, this new doctrine went through two phases of development, the first in the post–World War II period of decolonization, the second—an intensification of the diffusion effort—after the rise of socialist countries in the Third World, and particularly after the Cuban revolutionary victory in 1959.

A number of historical works appeared in the 1960s as part of this intellectual process. Their central purpose was to show that the European pattern of development, including most particularly the development of capitalism, had been somehow the one, the natural course of human progress, and many of these volumes quite explicitly drew the ideological conclusion that the proper, natural course of future development in the Third World would be to follow this natural European pattern (but not, of course, slavishly). The most influential of these works was Rostow's 1960 volume, *The Stages of Economic Growth: A Non-Communist Manifesto*. This book was a plain assertion that Europe's past formula for development, up to and including capitalism, was the only workable formula for non-Europe's future development. Rostow married world history to world development in a single diffusionist argument.[3]

But there was a problem. Precisely *what* had caused the unique rise of Europe? Now we must return for a moment to the classical diffusionist period. European historians of the prior century were unanimous (I think) in their acceptance of the fundamental idea that Europe has been naturally and uniquely progressive. For most of them the basic force underlying the process was unquestioned: some drew on their religious faith, others built on metaphysical ideas (like Hegel's evolving "spirit"), others appealed to a Smithian or Utilitarian idea of individual human activity and purpose, and still others invoked the natural environment, or demographic behavior, or class struggle, or something else, but I think it likely that all of them held a common conception of an underlying force of Progress, a basic directional force like the Solar Wind, in relation to which partial facts (economic, psychological, environmental, and so on) were epiphenomenal or merely symptomatic. After World War II, however, a profoundly different set of basic beliefs became the norm. Now the problem of explaining the rise of Europe was seen as a *total* problem. That is, all of the phenomenon, including its most fundamental dynamics, required explanation, or, stated differently, had to be put into an explicit model in which explicit variables or "factors" were identified. There were, doubtless, many reasons for this new—or rather newly popular—approach, among them the maturation of the discipline of history itself, the loss of faith (in this age of chaos) in the idea of inevitable progress, the general secularization of European thought, and the development of social science disciplines.[4] But whatever the overall explanation, what emerged was a set of historical models, some new, others (like Weberianism) refurbished, that explicitly tried to explain the "European miracle" in terms of specific causal factors. This was the signature of the modern form of diffusionist history. The effect of the modernization perspective was not, by any means, dominant in all of historical scholarship, but it was so in writings that sought to explain the larger transformations of European history, and particularly the problem of explaining the medieval changes that brought about the rise of capitalism and modernity—the "European miracle."[5] We will review many of these explanatory propositions later in the present chapter.

The Critique

The modernization approach was quickly challenged because it gave no real role to non-Europe, past and present, save as an essentially passive recipient of diffusions from Europe. For the period before 1492, it claimed that the significant evolutionary processes took place in Greater Europe. For the period from 1492 to World War II, it claimed that evolutionary processes continued to effloresce mainly in Europe and that colonialism brought the fruits of this progress to non-Europe. For the present and the future, progress for non-Europe (the Third World) would consist of the continued spread of innovations, mainly through mechanisms basically inherited from the colonial era. These propositions were distinctly unpopular among intellectuals of the Third World, and the rapid development of Third World scholarship in this postcolonial period led rather quickly to the emergence of a critical, even rejectionist, body of thought, including a new historiography.

The basic thinking went about as follows: For the precolonial era, it was necessary to resurrect one's own history and find out how it had contributed to the history of the world. (Colonialist history dismissed precolonial history for some colonies and distorted it for others. Therefore, as Amilcar Cabral said with deep irony, when the colonies gain their independence, they re-enter history.[6]) For the colonial era, the belief that colonialism itself the was source of all progress was patently untrue and colonial history had to be rewritten to show how it had led to poverty rather than to progress. On a world scale, new models had to be developed to show that colonialism, far from diffusing modernization to non-European societies, had diffused a mixture of good and bad innovations, which, for much of the world, had been a process not of development but of underdevelopment.[7] This body of thought came to be known as "underdevelopment theory" in Africa and Asia and as "dependency theory" in Latin America. Out of it came the first serious critique of Eurocentric historiography.

One can trace the origins of this critique to earlier writings by a small number of historians, most of them colonial subjects, often writing in exile. Their main theme was a documentation of the negative effects of colonialism on a particular place and people. Some of the writers—among them W. E. B. DuBois, R. Palme Dutt, K. M. Panikkar, M. N. Roy, J. C. Van Leur, C. L. R. James, George Padmore, and Eric Williams—developed arguments on colonial processes at a world scale, and on the role of colonialism in the postmedieval and modern rise of capitalism and Europe. James showed that Caribbean slaves had played a role in the development of capitalism not fundamentally different from that of the working class in Europe. Williams's work ultimately had the strongest impact on Eurocentric history. He showed that the wealth from slavery and the slave plantation had been the crucial factor in the amassing of capital for the Industrial Revolution in England. This argument was the first major demonstration that non-Europe had played a central role in modernization itself, and a sizeable literature has grown up among European historians arguing about, and generally trying to counter, what is now generally called "the Williams thesis."[8] During and after the period of decolonization this critical historical literature expanded greatly, and a large number of scholars began a direct attack on the diffusionist history of the colonial period.

Much of this work will be discussed in later chapters. Here I want to make several concrete points about the critique. First of all, a number of historians from the European world (Bernal, Frank, Wallerstein, and others) joined Third World historians as central figures in the movement. Second, while the critique focused a good deal of attention on the pre-1492 period, showing that development had taken place in non-Europe—for instance, Sharma and Habib documented the development of feudal and postfeudal society in medieval India—for a long time very little scholarly attention, within this critical tradition, was focused on pre-1492 Europe; the first major work that dealt with this problematic—from a non-Eurocentric point of view, that is—was Amin's 1974 volume *Accumulation on a World Scale*.[9] This is perhaps understandable since most of the critical historians were themselves from the Third World, not from Europe, and European history was not often a major interest. Yet it was anomalous nonetheless. The core of the modernization doctrine in history was, after all, the argument that Europe had begun to modernize before other regions and before it established colonial control of other regions. Thus, to refute the basic thesis one would have to show (as I try to now in this book) that pre-1492 Europe was not uniquely progressive. Of course, part of this argument consists of demonstrations that other regions *were* progressive. But part of it must consist of refutations of the various "miracle" propositions, those that claim to find in ancient or medieval Europe some special quality of progressiveness.

There is another, quite curious anomaly in the relative lack of attention to pre-1492 Europe by historians in this critical tradition, the tradition associated with underdevelopment or dependency theory and the critique of colonialism. This has to do with the curious relation between Third World scholarship, much of which is Marxist, and the Marxist scholarship of the European world. European Marxists were among the main critics of colonialism and among the main contributors to dependency–underdevelopment theory. Marxist theory also inherited from former times a strong anticolonial flavor and a profound skepticism regarding the theories propounded by nineteenth-century mainstream European historians.[10] Yet, oddly, most European Marxist historians writing about pre-1492 Europe have tended to argue in *favor* of the uniqueness-of-Europe doctrine.

Mainstream European historians have also contributed to the critique of the central Eurocentric doctrine. This is not an anomaly. Scholars try to pursue the truth and accept it whether or not it accords with their ideological or cultural preferences. To some extent they succeed. Thus a number of European scholars specializing in non-Europe have uncovered some of the most important evidence against the Eurocentric model of pre-1492 world history. The work of the Dutch historian J. C. Van Leur in the 1930s, concerning the economic history of South and Southeast Asia is a classic example of this antisystemic scholarship.[11] Another example relates to Chinese history. Half a century ago Duyvendak uncovered truly crucial facts about China's long-range voyaging in medieval times. Later, Needham and his associates produced a series of studies about the history of Chinese science and technology that had a profound impact on the Eurocentric model and (as we will see) forced Eurocentric historians to abandon a large piece of their argument concerning the supposed uniqueness of medieval European technology. Other Western scholars, like Wheatley and Elvin, delving into the empirical history of China with indifference to ideological questions, have produced other sorts of evidence about China's progressiveness in ancient and medieval times.[12] All of this damages the miracle theory, although, as we will see later in this chapter, Eurocentric historians have found ways to repair most of the damage.

The critique of Eurocentric history is a very large subject, and our concern in this volume is with just one part of it: the critique of European miracle theories about the world before 1492, and related theories which treat early modern world history as though non-Europe, and colonialism, were merely marginal to evolutionary processes. On these issues the critique has not progressed very far. I will give a few examples of important recent contributions, and others will be cited throughout this book. Janet Abu-Lughod's recent (1989) book, *Before European Hegemony: The World System A.D. 1250–1350*, is a seminal study that demonstrates (I think conclusively) that Europe was not more progressive and not more advanced than other civilizations in A.D. 1350. Having made this demonstration, she offers only a tentative and partial explanation for the selective rise of Europe, and decline of the Orient, *after* 1350. She suggests that the divergence took place in the period between 1350 and 1492. (I argue in this book that the divergence occured only after 1492, with the beginnings of massive accumulation in the Western Hemisphere, a windfall that did not accrue to non-European civilizations of the Eastern Hemisphere, and so gave Europeans their first and decisive advantage over these other civilizations. See Chapter 4 below.) Samir Amin has argued in various recent works that Europe was not more advanced than Africa and Asia at the end of the Middle Ages, but rather was more unstable: because of its marginal location at the edge of the hemispheric zone of civilization, medieval class society was less fully seated, less stable, less indurated in Europe than elsewhere, and so Europe changed toward capitalism more readily.[13] This argument, although it does not grant any "miracle" to pre-1492 Europe, nonetheless allows one of the old beliefs to stand: that Europe was more dynamic than non-Europe during the Middle Ages. Martin Bernal's new book *Black Athena* appears to have little connection to the subject of the present volume, yet his arguments are very closely connected indeed. Bernal shows that European historians have created a myth about ancient Europe according to which African and Asian origins and innovations are written out of history: the goddess Athena was African. Bernal's work undercuts the still very popular theories about ancient Europe's supposed uniqueness, and also exposes the ethnocentric and ideological roots of much of the European scholarship that underlies the classical diffusionist model.[14] Edward Said's 1979 volume, *Orientalism*, a seminal critique of this process by which Eurocentrism and conservative ideology has dominated European scholarly writing about the Near East and Asia, is also important and quite relevant to our argument here. Other such works will be referred to as we proceed.[15]

The Countercritique

In recent years there has been an outpouring of writings that strongly defend the traditional Eurocentric view, upholding the European miracle theory in some of its various forms: we discuss many of these writings in this chapter. In the same stream of writings there are counterattacks against the more specific theories that question the traditional European views about slavery, colonialism, and the like (see Chapter 4), views that treat these processes and events as marginal in social evolution. And new theories (or modified forms of old theories) about the precise reason for the uniqueness of Europe are being put forward and discussed. I have a hunch that this is a scholarly movement that resonates with the new political attitudes concerning the Third World. In any event, the decade of the 1980s saw a number of writings of this sort and they appear to embody a rather conscious counterattack against the critical history discussed above.[16]

Some of the writers in this new literature are very self-consciously engaged in such a counterattack. A number of them are Marxists and are insisting that the true, the original, the *correct* Marxist doctrine recognizes the priority, past and present, of Europe. Robert Brenner, for example, boldly argues that capitalism was invented by northwestern Europeans, with no help from others, and therefore (600 years later) we must acknowledge the continued priority of Europe. A number of other Marxists, like Perry Anderson and Bill Warren, argue similar positions.[17] Among mainstream historians the most dramatic event was the appearance in 1981 of Eric L. Jones's *The European Miracle*. This book is a remarkable recital of a goodly share of the colonial-era ideas about the precocity of Europe and the backwardness and irrationality of non-Europe. More remarkable still is the positive reception this book has received among many scholars, as though most of these old doctrines had not long since been disproved.

Another movement at present is an attempt to find qualities present in ancient and medieval European culture, and absent in other cultures, which were the reasons for European development: qualities in the European family, the European political system, the European mind, and so forth. This movement is actively resurrecting the turn-of-the-century views of Max Weber about Europe's supposed "rationality" and the like; indeed, most (not all) of these scholars can be thought of as Weberians and many of them define themselves in that way. I will discuss Weber's views later in this chapter, along with the views of some modern Weberian scholars, among them Michael Mann and John A. Hall.

In the next section of this chapter I will try to extract the most important of these newer views proclaiming Europe's pre-1492 "miracle," and I will try to show that these views are mistaken.

THE MYTH

The myth of the European miracle is the doctrine that the rise of Europe resulted, essentially, from historical forces generated within Europe itself; that Europe's rise above other civilizations, in terms of level of development or rate of development or both, began before the dawn of the modern era, before 1492; that the post-1492 modernization of Europe came about essentially because of the working out of these older internal forces, not because of the inflowing of wealth and innovations from non-Europe; and that the post-1492 history of the non-European (colonial) world was essentially an outflowing of modernization from Europe. The core of the myth is the set of arguments about ancient and medieval Europe that allow the claim to be made, as truth, that Europe in 1492 was more modernized, or was modernizing more rapidly, than the rest of the world.

This is a myth in the classical sense of the word: a story about the rise of a culture that is believed widely by the members of that culture. It is also a myth in the sense of the word that implies something not true. In the following discussion I will unravel the fabric of this myth and show that the strands of belief that compose it are very feeble.[18]

The number of distinguishable belief statements that make up this myth are, I am sure, uncountable. One of the many reasons the myth is so durable is the fact that the basic generalization, the doctrine of the miracle, is supported by such a great variety of individual beliefs that historians of a given era can disprove some subset of these beliefs and yet the supporters of the myth can merely shift to other beliefs as grounding for the myth.

A more fundamental problem has to do with the way beliefs are licensed. Beliefs tend to gain acceptance if they support the myth, and are either rejected or denied attention if they do not do so. One part of this problem of belief licensing (and relicensing, delicensing, etc.) poses a particularly, perhaps uniquely, serious difficulty for efforts to critique the miracle theory. Many of the beliefs that support this theory are *implicit*, not *explicit*; that is, they do not enter into the scholarly discourse of historians, and sometimes they do not enter even into conscious discourse in general. (Recall the discussion of implicit beliefs in Chapter 1.) Many of these beliefs we learn as children. Others seem self-evidently "reasonable" because they accord with deep values of the culture, or with other, accepted beliefs (historical, practical, religious, and so on). Thus, the conviction that ancient and medieval Europe was more progressive than other civilizations is supported by explicit beliefs, but these lie in a matrix of implicit beliefs—unquestioned and usually unnoticed—about the progressive Europeans who "were our ancestors." By contrast, the matrix of implicit beliefs about historical non-Europe includes ideas of alienness, savagery, cruelty, cannibalism, deceitfulness, stupidity, cupidity, immodesty, dirtiness, disease, and so on—a matrix firmly supporting the general belief that non-Europe *cannot* have been progressive. Examples of these sorts of implicit beliefs, both positive and negative, will appear as we proceed.

One kind of explicit belief about European superiority will not be discussed here in detail. This is the openly religious statement, grounded in faith, that a Christian god will naturally raise His people higher than all others. Although we will refer to this view in various contexts, it is not the kind of argument that can be analyzed or criticized, because it is grounded in a faith that cannot be tested empirically; some believe it to be true and others do not, and that is as far as the matter can be taken. Suffice it to say at this point: the religious argument was so nearly universally accepted down to the nineteenth century that other arguments were not seen as necessary to many European intellectuals. Why, indeed, ask for the reasons that Christian Europeans are superior when we know that unbelievers will not go to heaven and, in this world, will not enjoy the grace of God? Unbelievers will naturally be rendered less intelligent, less fortunate, and so on. So long as scholars and educated people believed that religion underlies all things, including science, and that God intervenes to control human affairs, then it could simply be assumed that Europeans were superior because that was God's will. It is what you would expect a Christian god to do for Christians, particularly for those Christians who worship Him in the right way. By the middle of the nineteenth century, when Eurocentric diffusionism was at its height, when Europeans hardly ever doubted their superiority over everyone else, although the explicitly religious arguments were disappearing from scholarly discourse, they were still there implicitly, as implicit beliefs. I rather suspect that the great majority of arguments for the superiority of Europeans were finally grounded in a religious faith: if the European environment is superior, that is because God made it so; if the white race is superior, that is because God made it so; if Europeans are more rational than everyone else and this has *no* explicit explanation, one can infer that it is the work of God; and so on. I do not know to what extent this kind of implicit appeal to the Deity is still present but unnoticed in the thinking of contemporary Eurocentric scholars, but I am certain that it quite often is so. We will from time to time discuss the ideas of scholars (like Lynn White, Jr. and K. F. Werner) who explicitly connect their views about history with their religious beliefs; in such thinkers the directionality of the causal arguments will be noted but not in any sense condemned. Sometimes a scholar makes such arguments unconsciously and implicitly. The only really disturbing cases are those that exhibit conscious hypocrisy.

Biology

Basically two sorts of argument have been used to explain the uniqueness, the superiority, of Europe and Europeans. One sort appeals to some noncultural force or factor as prime cause; the other finds the prime cause within culture itself. Setting aside the doctrine of divine intervention (which one would generally think of as a cause external to human culture, although the point is theologically rather complicated), two kinds of noncultural, external causation are common. One appeals to human biology, the other to the natural environment.

Race

Biological arguments assert, in general, that Europeans are superior, biologically, to non-Europeans. The classical and typical form of this argument was biological racism, the idea that Europeans had superior heredity, and so were born with abilities greater than those displayed by non-Europeans. Europeans were brighter, better, and bolder than non-Europeans because of their heredity. Generally the descriptive category used was not "Europeans" but rather "members of the white race," but the distinction was not usually very important. Non-Europeans who were classified as members of the white (so-called) race were nonetheless believed to be inferior because they belonged to inferior subraces. Sometimes Europeans themselves were divided into superior and inferior subraces. Early in the nineteenth century it was widely, though not dominantly, believed that white people were not even of the same biological species as people of other races. This theory, "polygenesis," claimed to have both biblical and scientific support.[19] Its primary importance was its use as a rationalization for slavery: if Africans were not truly human, why, enslaving them could not be an evil act. This theory melted away during the course of the nineteenth century, mainly because it was offensive to liberal, modern, antislavery thought; what replaced it, however, was not much of an improvement. This was the doctrine we call classical racism, the belief that different human races have different endowments, just as different breeds of domestic animals have such differences—differences of intelligence, aggressiveness, courage, and the rest—and these different endowments are matters of biological inheritance. It was then argued, for instance, that Africans were endowed with lower intelligence than Europeans, so it was both natural and moral for Europeans to colonize Africa and make all decisions on behalf of the Africans who, themselves, either did not have the innate ability to make these decisions, to govern themselves, or were just sufficiently less intelligent than the Europeans that a period of European control and tutelage was necessary while these slow-thinking Africans learned how to govern themselves. Racism, in a word, had as its main function the justification of colonialism and all other forms of oppression visited upon non-Europeans, including minority peoples in countries such as the United States.

Toward the end of the nineteenth century racism acquired a pseudoscientific aura of apparent truth. Scientists claimed to have proof of the differences among races, particularly as to intelligence. Now, also, they were armed with Mendelian genetics, so it seemed eminently reasonable for scientists who were, themselves, racists, now to assert that they had proof of the truth of the racist theories in which they already believed. These so-called proofs were unmasked as pseudoscience in a slow process that continued down to World War II: indeed, even today there are more than a few so-called "scientific racists." What we need to notice about scientific racism is that it merely provided a new way of justifying something that was already almost universally believed to be true among Europeans. This meant, on the one hand, that it did not do much to intensify racism— which was in fact reaching a crescendo of intellectual and scholarly importance around the turn of the century for social reasons, having to do mainly with the growing importance of colonialism—and, on the other hand, that the disproving of scientific racism did not have much to do with the decline of racism's popularity in the present century. Racism emerged from prescientific roots and survived so long as it was useful, science or no science.

This is not to deny that scientific racism gave strong impetus to racist beliefs in society; in a time when science was acquiring great prestige and influence, scientific arguments were indeed important.[20] And scientific racism helped in some special ways. One important example here has to do with supposed differences in endowment among the supposed subraces within the so-called white race. Anti-Semitism, along with anti-Muslim attitudes relating to colonialism in the Middle East, led to a proliferation of theories about the inferiority of the so-called "Semitic" subrace. ("Semite," of course, should refer only to someone who speaks a Semitic language, such as Hebrew or Arabic.) This racist underpinning for anti-Semitism, then, became extremely useful for the general argument about the superiority of Europeans in history over *all* non-Europeans.[21] Semitic peoples were acknowledged to belong to the white race. Now it could be argued that only Europeans belonged to the really *superior* subrace of whites. Semites were inferior. So were the non-Semitic peoples of western Asia (Persians, Turks, and so forth). Indeed, in certain situations it was even argued that scientific racism proved the inferiority of southern and eastern Europeans: the northern, British and Germanic stock was truly superior to Italians, Slavs, and the rest. One famous instance of this sort of argument was the series of pseudoscientific testimonials given before the U.S. Congress at the time it was debating the first important immigration legislation in the 1920s. Scientists solemnly assured the Congress that southern Europeans were inferior people, and so should be excluded from free immigration to the United States in order to maintain the high racial quality of American stock.[22]

Today very few educated Europeans believe that there are genetic, inherited differences among the races as to intelligence or any other quality that might favor or inhibit social progress. Although a few people believe in the doctrine of classical racism, most of them are careful to keep their views to themselves, because today the doctrine is so thoroughly rejected, and viewed with such repugnance. If this were a book dealing with the history of ideas, we could go into the explanation for this transformation—the decline and fall of classical racism over a period of no more than two scholarly generations. During the 1920s the belief in the influence, major or minor, of inherited or racial differences was very widespread. After 1945 the theory was very rarely defended. Probably the key factor was Nazism. The Nazis grounded their ideology in this belief, claiming that the so-called Nordics were a Master Race, that inferior kinds of Europeans, inferior (so-called) subraces of whites—like the so-called Semitic race—and all other races as a whole deserved to be ruled by the Master Race. The Master Race, moreover, had the right to eliminate inferior subraces by genocide. Racism therefore was seen, and still is seen, as a component of the horrible ideology of Nazism. It is true that a small and fanatic group still preaches classical racism, and that a very small number of academics still proclaim its validity. The need to fight against the doctrine has not entirely ended. But this no longer is the important issue when we talk about Eurocentric prejudices, because another doctrine has largely replaced classical racism and performs much the same function, rooting its argument not in genetics but in culture. This doctrine can be thought of as cultural racism.[23]

Classical racism was so pervasive, down through the 1920s, that it probably figured as an explicit or implicit foundation for most arguments about the superiority of Europeans in history. It is as though the scholars who asked why Europe had risen and other societies had not done so had part of their answer at the outset: Europeans began with a genetic advantage, large or small, and then, throughout later history, they were always to a greater or lesser extent favored by the influence of their genetic superiority in matters intellectual, giving them superior decision-making ability, inventive ability, and so on. Here was an important reason why it did not, in those times, seem crucial to ask the question, "Why?" At root, the answer was considered self-evident. Even moderate racists, such as Max Weber (whose views we examine below), could therefore assume that European superiority was carried along by a kind of subtle undertow of genetics, that this made Europeans at all times slightly more "rational" (Weber's favorite word), and, therefore, slightly more progressive. It is very interesting that modern scholars retain the notion of the superiority of European "rationality," a notion derived from Weber and through him from all of nineteenth-century social thought, yet they vigorously deny that the source of this "rationality" is racial superiority. The intellectual contortions they have to go through to retain the one without the other will claim our interest when we discuss this matter.

Among the historians whose theories about the "European miracle" are at present widely supported, only Max Weber uses classical racist argument in a clear and overt way. But Weber was writing in the late nineteenth and early twentieth centuries, when classical racism was accepted by the majority of European scholars. Weber's arguments are a rather mild form of racism. He did indeed write of the "hereditary . . . hysteria of the Indian," a basis for his argument that Indian religion prevented Indian development.[24] He considered Africans to be genetically incapable of factory work.[25] Chinese have a "slowness in reacting to unusual [intellectual] stimuli," a "credulity" or "docility" that Weber thinks are wholly or partly hereditary traits.[26] And Europe's greater "rationality" has a definite hereditary basis.[27] But the fact remains that Weber gave much greater weight to nonracial factors. And Weber, for his time, was only a moderate racist.

Moderate racism is, today, a more serious problem in the world of scholars than is classical racism, because it is mainly an *implicit* theory. We noted that Weber believed in the significance of racial differences, but he referred to the matter very rarely; yet it must have been an unstated, perhaps implicit, part of many of his arguments about social evolution in general and the comparative evolution of Europe and non-Europe in particular. This was fairly typical of early twentieth-century scholarship, hence the racism in that scholarship often is too difficult to identify, and the arguments presented by scholars seem not to be racist when they do not explicitly mention race as a factor. But more serious still is the surviving influence of what I call *very* moderate racism. A great number, perhaps the majority, of mainstream scholars, in the period, say, of the 1920s, believed that racial differences were very slight and that the individual human being's capabilities and potentialities would not be predictable from his or her race, that race differences only appeared influential on a statistical basis for large groups: for instance, a slightly higher average "intelligence quotient" for whites as against blacks. This belief was consistent with militant opposition to racial discrimination. But it was not much better than classical racism when applied to questions of social evolution and comparison between European and non-European history. This is so because the historical arguments did not need to postulate *large* racial differences. If whites, on the average, held a tiny advantage over non-whites in, let us say, inventiveness, that tiny advantage, working out its influence over the centuries and millennia, would produce the result that whites built high civilization and nonwhites did not. In a sense, this *very* moderate racism was a more serious problem than ordinary racism, because it allowed scholars to take liberal positions in opposition to overt racial discrimination yet continue to believe that whites are superior genetically to nonwhites within the subject matter of their own fields—anthropology, geography, history, and so on. It did reduce the significance of race to that of one "factor" among many, but it did not eliminate racial explanations in the matter of Europe's supposed superiority.[28]

Probably it is unnecessary to say here that there is no credible evidence in support of the idea that races differ in genetic inheritance except in trivial matters like skin color. Even the idea of race is a vague abstraction and not really useful. For our purposes, the generalization that counts is this: racial differences explain *nothing* about culture or cultural evolution.[29]

Demography

Demographic behavior is usually considered the second important biological factor in explanations for the European miracle. Causality here is very murky. The specific form of theory with which I am concerned claims that Europeans, historically, controlled their population growth, whereas other peoples did not, so that Europe tended not to suffer overpopulation, and did not encounter what many historians call the "Malthusian disasters" that supposedly prevented forward progress in non-European societies. Malthusianism postulates, in its essence, that ordinary people do not control their sexual urges and so have more offspring than can be fed; a disaster of some sort ensues, with famine, pestilence, or war now reducing the population; and then people again breed more children than they should, and the cycle renews. Malthus considered this uncontrollable sexual urge to be general in the human species, although the educated classes could control themselves to some degree. Modern explanations for the "European miracle" modify this theory in a key respect: they assert that Europeans, historically, have had a cultural pattern of limiting the number of offspring whereas non-Europeans lacked this pattern; as a result, European population was maintained, throughout history, in rough long-term proportion to resources, in spite of periodic Malthusian crises, crises which were important in explaining various facts of European history but were vastly less significant than the supposed permanent grip of Malthusian forces—lack of demographic control and perennial overpopulation and misery—in non-European societies. Moreover, according to these historians, whenever technological advance or some other fortunate circumstance led to a rise in living standards, Europeans, unlike non-Europeans, did not allow their population to rise and thus eliminate the fruits of progress. We need to notice that this argument does not really center on the biology of reproduction but, again, on rationality: it is claimed or assumed that Europeans *think* about the problem of population and others do not. So demographic arguments for the superiority of Europe in history always return either to culture or to race: Europeans are more rational either because they have superior heredity or because they have superior culture. If non-Europeans have uncontrollable sexual urges, if they behave like the beasts in the field, this is either a sign of genetic inferiority or a cultural quirk. In the arguments most frequently encountered today, either culture is invoked or the matter is left ambiguous. Yet there is a paradox. Many historians believe that demography is an *independent* causal force in history. This belief is generally grounded in a quasi-biological argument: ordinary people have only partial success in controlling their sexual urges, and only partial success (or none at all) in controlling and spacing the birth of children.[30] Many supporters of the miracle myth simply argue, then, that Europeans have greater, though still partial, success in controlling their behavior than do non-Europeans. For instance, Eric L. Jones claims that this is what he calls "the quality of Europeanness":

> Europe did not spend the gifts of its environment as rapidly as it got them in a mere insensate multiplication of the common life.[31]

In other words, Europeans do not simply reproduce thoughtlessly, as non-Europeans do. Jones makes the same point in many different ways; for instance, he says of Chinese peasants that they preferred "breeding new people" to improving their economic and political circumstances.[32] John A. Hall makes a similar point:

> The expansion of the European economy did not occur [through expansion of cultivated acreage], as in late traditional China, because improvements in output were not eaten up by a massive growth in population. The ratio between population and [cultivated] acreage in Europe remained favourable ultimately because of the *relative continence of the European family*.[33]

Europeans are sexually "continent." Europeans, therefore, do not suffer overpopulation.

Another point worthy of notice: the historians have no difficulty invoking population growth as a *positive* factor in European history, as an indication of progress—for instance, the growth of Europe's population in the eleventh century and thereafter is seen as proof that medieval Europe was healthily advancing—whereas population growth in non-European societies is seen as *negative*, as the working out of Malthusian laws of "overpopulation." We will encounter examples of this very one-sided form of argument as we proceed.

There is now abundant evidence that all societies practice population control.[34] They seem to do so very effectively at the aggregate level: although individual family groups may or may not succeed in controlling the incidence of conception—because the methods of birth control are often rather hit or miss—the society as a whole seems able to encourage population growth or discourage it quite successfully.[35] It is most unlikely that population growth takes place when the society would, in the long run, suffer as a consequence; or, to put the matter more precisely, the members of a society change their demographic behavior in the space of a couple of generations when it becomes clear to them, as they observe changes around them in the probability of infants surviving to adulthood, and so on, that such change is desirable. The knowledge that all of this is indeed the case has only appeared recently.[36] But this accumulating evidence really cuts the ground from under all Malthusian theories, historical and contemporary, European and non-European. The argument that population growth is an automatic, biological process, which occurs whether or not there is food enough to feed additional mouths, is simply wrong.[37]

When the evidence began to accumulate that historic European populations did indeed practice population control, some scholars decided, appropriately, that the old Malthusian models do not make sense in Europe: family size was kept small, births were controlled, and so forth, back in the Middle Ages. This called for rejection of one traditional theory, the idea that there is a general pattern for all so-called preindustrial societies (or "traditional societies," or "peasant societies"), involving high and uncontrolled birth rates, large families, etc., and thus, inevitably, a Malthusian trend toward overpopulation. Most historians today seem still to be wedded to the Malthusian theory, and still believe that medieval social change in Europe was mainly, or at least partly, produced by Malthusian cycles of overpopulation.[38] But some historians now reject this perspective: population, in essence, is viewed by them as a dependent variable, not an independent variable.

But only in Europe. This newer, essentially anti-Malthusian, theory was rapidly shaped into a new explanation for the European "miracle." It was simply asserted that *European* people, historically, controlled their population, and so, when they accumulated surpluses of food and commodities and wealth, these were not subsequently dissipated as population grew out of control and stole away the savings. This pattern, said these historians, was uniquely European and it was one important reason why Europeans were able to accumulate wealth and eventually rise into modernity and capitalism. Non-Europeans, by contrast, continued in their unmodern, traditional pattern, with overpopulation wiping out all benefits of forward progress. These Eurocentric historians, in a manner characteristic of tunnel history, did not notice that new scholarship on demography, some of it on historical demography—in various non-European parts of the world, from India to Barbados—was tending to overturn Malthusian notions as they apply to non-Europe just as readily as had been the case a decade earlier in European scholarship.[39]

We will take up a related matter, theories about the uniqueness of the European family, at a later point in this chapter. For now, it is enough to say, simply, that the demographic arguments in support of the European miracle are utterly unconvincing. This holds true both for the theories grounded in the belief in baleful Malthusian forces and for those which assert that Europeans, uniquely, know how to avoid those forces.

Environment

Environmental determinism, the theory that the natural environment strongly influences human affairs and human history, is no longer a popular doctrine, but it is still used quite regularly in explanations for the European "miracle." The point should, however, be qualified: Environmental determinism, in the form it is used in these arguments today, is a "determinism" in a limited sense only. It does not claim that the environment explains everything, or that the environment is the most important explanatory factor. It is deterministic in the sense that it treats the environment as a separate, simple cause or "factor" not mediated by culture: something external to culture and influencing it from the outside. One finds each Eurocentric historian adding one or a number of environmentalistic arguments or factors to the mix, which, as a whole, explains the superiority of Europe. Even the historians who want to build their case mainly on social or political or intellectual foundations nonetheless manage (I know of very few exceptions) to throw one or more environmentalistic arguments into the stewpot, to give it body or flavor.

Environmentalistic arguments can be stacked into two piles. One consists of the set of claims about the superior qualities of Europe's environment and how they help to explain the rise of Europe. The other pile consists of arguments as to why the nasty environments of other places have blocked development there. The two sorts of argument are—perhaps surprisingly—very different in form. Let us begin with the latter.

Two classical environmentalistic theories have been used over and over again, and still remain in use, to explain the (supposed) nondevelopment of Africa and Asia. The first theory argues that tropical regions are innately inferior to cooler regions. This theory is used, for the most part, to dispose of Africa. The second theory argues that peoples in arid regions are held back from development because aridity necessitates irrigation, and irrigation along with related features of irrigated river-valley life leads, again necessarily, to the kind of civilization that is historically stagnant. This theory disposes of Asian civilizations, along with Egypt. (Most of Asia is *not*, in fact, arid.) I will discuss both of these theories, and more briefly a few other environmentalistic explanations for the backwardness of Africa and Asia.

Nasty-Tropical Africa

The idea that tropical climates are nasty, and inhibit the forward march of civilization, is a very old one in European thought.[40] During the nineteenth century this notion was widely used to show why Africans have (supposedly) remained uncivilized, and must naturally accept European colonial control; it was one of the core theories of classical diffusionism.[41] It was routinely built in to theories about the uniqueness of Europe, the European miracle, although I must add that historians did not usually consider the reasons for Africa's nonrise worth bothering about: the matter was considered self-explanatory, and labors were devoted to the seemingly more important question why Asia (and North Africa) did not rise. And the tropical-nastiness theory is still quite regularly employed in historians' arguments. It is, for instance, crucial to Jones's argument in *The European Miracle*. It is also significant in a special arena: the debates among historians about African slavery, the slave trade, and the slave plantation system. By contrast, geographers, who in general know something about the natural environment and have long grappled with the theories of environmental determinism, do not, today, take the nasty-tropics theory very seriously.[42]

The tropical-nastiness doctrine consists mainly of three distinct theories. The first concerns itself with the supposed negative effect of a hot, humid climate on the human mind and body; the second, with the supposed inferiority of tropical climates for food production; the third, with the supposed prevalence of disease in tropical regions. Down to the 1940s, or thereabouts, there was some ambivalence among Europeans as to the thesis that humans cannot labor as effectively in the humid tropics as in other climates. The majority opinion had been that Africans can labor under the hot sun—a convenient rationalization for plantation slavery— but that Europeans cannot do so, although some diffusionist arguments were built on the idea that tropical conditions induce sloth, indolence, etc., in everyone, and thus the need for control-at-a-distance from temperate-climate civilizations. Eventually it became clear, from many sources of evidence including physiological studies, that human bodies of all sorts can labor as effectively in the tropics as elsewhere if the bodies in question have had time to adjust to tropical conditions.[43] Although the claim that people in humid tropical regions cannot *think* as well as people in temperate regions—the tropical sun "boils the brain"—was almost universally accepted by Europeans in the last century and was incorporated into some well-known theories about European civilizational superiority put forward (for instance by Huntington and Markham) in the first half of the present century, this theory, never grounded in evidence, is now rejected across the board. Eric L. Jones is one of the few present-day European historians who is unaware that this theory of "climatic energy" (as it was called in the old days) has been exploded. Not only does Jones accept it as valid but he gives it some considerable significance:

Civilizations had long been rising and falling in warm latitudes, although they appear to have been springing up farther and farther north. Such explanation as the literature offers for this shift is essentially climatic (Gilfillan 1920; Lambert 1971). On the one hand it correlates mean temperature and the output of human energy, and on the other it claims that in warm regions man was subject to the build-up of endoparasitic infestation which caused each society there to reach a plateau of attainment and then stagnate.[44]

It should be noted that "the literature" does not support any of this. The scholarly literature roundly rejects the theory of "climatic energy," which has not in fact been seriously defended since the 1950s. Climatic determinism as a whole is almost a dead letter. There has been no northward "shift" of civilization (ancient civilizations were located from the equator to 45° latitude). The notion that tropical regions are so parasite-infested that cultures stagnate is false (it will briefly claim our attention later). Jones's general belief that hot climates are "debilitating" has little if any scholarly support, and does not in turn support the idea that Europe's climate led to a European miracle.[45] There simply is no basis for arguing that midlatitude climates are superior to tropical climates in terms of psychological or physical effects on human beings.

The fruitfulness of tropical environments was much debated by nineteenth-century European scholars. Some argued that tropical regions are lush and bountiful, but used this proposition not as a basis for asserting high development potential but rather the contrary: tropical environments are too bountiful to offer what Arnold Toynbee might have called a sufficient challenge to humanity, and so progress did not take place except under colonial guidance. We deal with this thesis later in the chapter. Our concern now is with the exact opposite thesis, which asserts, quite simply, that tropical environments are miserably poor in their potential for agriculture and it is this that prevents tropical regions from developing. (The term "tropics," or more properly "humid tropics," refers to regions with no cold season and with moderate to high rainfall: roughly 750 mm. of rain or more per year. Most of sub-Saharan Africa falls within the humid tropics, as does most of southern and southeastern Asia and most of Middle and South America.) According to this thesis, little food can be produced on a given piece of land in the humid tropics.

Now this thesis—low agricultural productivity of tropical land—is hard to defend on the evidence, since population densities in the humid tropics range from very low (in parts of the Amazon Basin, for instance) to extraordinarily high (in Java, Bangladesh, El Salvador, Barbados, Rwanda, and many other regions). The argument is grounded, rather, in a theory about the nature of tropical soils. To put matters in perspective, the scientific study of tropical soils is a very young field; few contributions of any consequence can be dated earlier than World War II if we exclude highly particular studies about the soils used for plantation crops like sugar-cane.[46] The really crucial information, relevant to the proof or disproof of the traditional theories, was obtained in research carried out in a number of experimental centers in the late 1940s and thereafter, and then slowly disseminated to the world scholarly community. The result is that some historians (including some current writers about the "miracle") can still, today, make use of quite fantastic theories about the supposedly evil nature of tropical soils. Some of these theories were refuted quite recently; others are now, today, very close to refutation; a few others, hardly defensible, nonetheless still lie around in the scholarly literature, unrefuted. A brief, only slightly technical, comment on these theories is necessary at this point of the discussion.

The old standard theory about tropical soils runs about as follows: Because of the high heat and abundant rainfall in the humid tropics, these soils cannot accumulate organic topsoil, since organic matter decomposes rapidly and is quickly leached out as rainwater seeps downward in the soil. Tropical soils, therefore, are low in plant nutrients. Also, they are subject to severe erosion, partly because tropical landforms tend to have high average slopes, partly because high rainfall means high runoff and therefore much surface erosion.

These two basic physical propositions, about infertility of the soil and erosion respectively, were then married to a cultural proposition, in the following argument. Because of the low fertility and erodibility of tropical soils, farmers must practice what is called "shifting agriculture." (This is a farming system in which a field is prepared by clearing a piece of forest with the use of fire, and then, after one or more years of cultivation, is abandoned, another field being then cleared in turn and in its turn abandoned, in a continuous process of shifting fields.) The standard theory (as it was argued before the 1950s or thereabouts) then makes a series of sweeping generalizations about the combined effects of infertile, erodible tropical soils and shifting agriculture. It was claimed that farmers cannot return to the original piece of land after it has been abandoned and left fallow for some time, since the soils are too poor to regenerate and since burning the piece of land permanently damages the soil. Even where the environment is lush enough so that farmers can return and re-use each field, the production from each field will be less and less as the cycles of use and abandonment continue, and the soil will become more and more infertile until finally it is unusable. All of this implied that peasant communities would be unable to remain in any one region permanently;

that the villages themselves would have to be moved as large expanses of forest land were used and then abandoned and the people needed to move elsewhere to find fresh, cultivable lands. The result would be grave indeed: a very sparse and highly mobile population with little or no chance of developing large-scale trade, cities, and stable states.

The entire argument was inserted at this point into world history and world geography. Most of sub-Saharan Africa is tropical. Therefore, shifting agriculture must be used there and no civilization can develop there. Or if civilization manages somehow to evolve to one extent or another, it must sooner or later collapse. (The decline of the classical lowland Mayan civilization was regularly used as the type example of this historical outcome; it is still so used by a few scholars.)

This traditional model of tropical soils and shifting agriculture was gradually softened as more information became available to students of tropical soils and tropical agriculture. By the 1960s it was generally known to the specialist community (though not yet to the majority of historians and social scientists) that shifting agriculture does not damage the soil under normal—that is, typical and widespread—circumstances.[47] Almost never do shifting cultivators move their villages because of soil exhaustion. (It became known, also, that shifting cultivation had been widespread for many centuries in Europe, and there it had not apparently done any damage to the environment.) From this emerged a softer cultural model: *if* population densities are low, so that farmers can leave each abandoned field for the many years required for soil and vegetation regeneration, *then* shifting agriculture will remain an equilibrium farming system, and there will be no long-term deterioration of the environment.

But even this modification would not much alter the historical judgment about Africa and some other tropical regions: historians might still believe that any civilization arising in such a tropical area cannot be very complex under ordinary circumstances, because a low population density of food producers would not seem to provide the basis for a substantial complex of urban centers, religious centers, states, and the like. (As to the lowland Mayan civilization, some scholars continued to claim that shifting agriculture had destroyed the subsistence base of that civilization, while others claimed that the decline and abandonment of cultural centers like Tikal had been due to other sorts of processes.) In any event, the prevailingly Malthusian view of peasants led to the general assumption that population would grow out of control and regions of tropical shifting agriculture would never rise very high in places like tropical Africa, anyway.

The model in this form is still current among historians. It is common in the European miracle literature, as the primary environmental basis for claiming that Africa could not have "risen" in the same way that Europe did. It is explicit in the work of some non-African Africanist historians.[48]

Evidence is now available to reject this entire theory, the notion that tropical soils are bad for agriculture and therefore somehow inhibit human history. First of all, we now know that tropical soils are not *inferior*. They are *different*. Because of the higher rate of chemical and physical weathering under humid-tropical conditions, soil production from the underlying rock is much more rapid than is the case in cooler climates. Therefore, soils maintain their fertility in considerable degree from the dissolution of minerals, and much less from the accumulation of organic matter. Erosion tends to be more serious, but regeneration tends to be more rapid. Tropical soils that have developed on rocks that are rich in plant-nutrient minerals are exceptionally *fertile*. Those developed on rocks that are not nutrient-rich are exceptionally *infertile*. There is no basis for comparing tropical and temperate "averages": neither is better; they are different.

Evidence is also available now to reject the old view of shifting agriculture. Farmers practice it on poor soils, being careful not to let fires get out of control. They use a great number of techniques to assist the natural vegetation to regenerate, and to increase soil fertility in the cultivated fields, including green and animal manuring. When there is a land shortage, the shifting rotation is shortened, by the use of various techniques of intensification such as mounding or terracing, increased labor applied to such matters as weeding, adoption of different crops and different varieties, and many more.[49] The correlation between shifting agriculture and low population density is a function of history, not ecology. In the Americas it reflects in part the post-Columbian depopulation. (The Amazon Basin, which today may have a farming population of perhaps one million, probably had seven times that number in 1492.[50]) In part also it reflects the often unnoticed fact that in most regions of the American tropics giant cattle ranches have pushed farmers off the better land, giving a statistical appearance of low population density: cattle have replaced people.[51] The same process occurred in white-settled regions of southern Africa. Elsewhere in tropical Africa, most farmers practice forms of agriculture that should not be described as shifting agriculture except in marginal regions, like mountainsides and semi-arid wastes.[52] Typically, they are sedentary or semisedentary farming systems, involving such things as semipermanent yam mounds and tree-crop farming, or irrigated farming, or mixed farming, or they are systems in which the period of cultivation exceeds the period of fallow, and fertility is maintained by many different cultural practices including the use of green and animal manure. But the most important generalization is this simple one: where there *is* soil degradation, and hunger, and poor farming, it reflects cultural causes from recent history or colonial history. It does not reflect inherent limitations of tropical agriculture and it does not reflect technological ignorance on the part of farmers.

A few variant theories about the low food production potential of the humid-tropical in general and Africa in particular are invoked by some historians. Some claim that Africans were unable even to farm in Africa's humid-tropical regions until new technology, invented by *non*-Africans, diffused into the continent. One form of this argument claims that ironworking was brought into Africa a little over 2,000 years ago, perhaps by the Romans, and only then were Africans able to tackle the tropical forests.[53] Since shifting agriculture was practiced with stone implements in ancient Europe, and since ironworking appeared in Africa—possibly after independent invention—around 800 B.C. or earlier, this theory is invalid.

An even more outlandish theory builds on the fact that some Southeast Asian crops diffused into East Africa two millennia ago. Some historians blandly assert that Africans could not farm in the tropical forests until these tropical crops, domesticated by *non*-Africans, became available for planting in the African forests.[54] It has, in fact, been well known for a long time that Africans domesticated a great number of crops for humid-tropical regions, notably many varieties of the yam (*Dioscorea* spp.) the prime staple food there.[55] Both of these diffusionist myths seem to be connected to an important colonialist belief—now important as an excuse for *apartheid* in South Africa. The belief is that advanced African cultures, with agriculture, trade, and states, expanded southward through the continent only very late in history, mainly because of the forbidding nature of the tropical part of the continent which held back their southward movement. According to this myth of emptiness (see Chapter 1), they had not arrived in (most of) South Africa when the Europeans took over that region, and supporters of white supremacy in South Africa claim, on this basis, that whites, having arrived first, have political and economic rights to own the land of South Africa.[56] In fact, the expansion of agricultural peoples in Africa occurred thousands of years ago; recent archaeology has shown that rain forest regions were settled by farmers *at least* 3,000 years ago, and whites did *not* arrive first in South Africa.

Another variant of the myth of tropical nastiness as applied to Africa is the notion that rainfall variability in tropical Africa is uniquely and devastatingly high, such that agriculture could not, in precolonial times, be dependably practiced and famines were frequent and widespread. This then leads into various arguments about African backwardness. For some historians (among them Philip Curtin) there was a basic tendency toward mobility of populations and internal slave trading; for others (notoriously Joseph Miller), there was savagery and cannibalism.[57] These historians belong to what I will call the "absolutionist" school of Eurocentric Africanist history, which absolves Europeans of most of the responsibility for slavery and the problems of modern Africa by finding explanations for such matters within Africa itself, often in the African environment. The European miracle historians make regular use of these arguments for purposes of comparison. It happens to be the case that rainfall variability is a serious problem in all semi-arid regions, including the Sahel zone south of the Sahara in Africa and also the Great Plains in the central United States, the steppes of Russia, and so on. Africa is not unique. The more humid parts of this continent do not have a peculiar problem of climatic uncertainty: this is a historians' myth, partly traditional, partly resurrected after the recent Sahelian–Sudanic famines. The latter did not reflect drought-proneness of the region. They reflected human problems, mostly inherited from the colonial period, which became disasters when ordinary rainfall cycles went through their dry phases.[58]

Much the same sort of argument applies to other parts of the tropics. In southern and southeastern Asia we tend, broadly, to find permanent, that is, sedentary, agriculture on lands with relatively high fertility, and shifting agriculture or tree crops on lands with low fertility. The experience of colonialism during the past two centuries or so has muddled that picture somewhat, since there have been major population movements and large population increases in some areas—with consequent distortions of farming systems in reaction to land shortage and other nonenvironmental pressures. Yet the generalization still holds in regard to the environment: tropical conditions do not carry the implication of poor agricultural potential.[59]

What of the opposing theory, that the tropics are lush and bountiful? Down through the middle of the nineteenth century this theory was very widely accepted, and also this corollary: Since the fruits of the earth are so easily obtained in tropical climes, humans do not have to exert themselves to make a living. And so they do not progress. The argument was then woven into many different theories. Buckle put forward one such theory.[60] Marx put forward another, rather tersely, in a footnote in Volume 1 of *Capital*, asserting without discussion that tropical regions do not develop toward capitalism because here "Nature is too lavish . . . She does not impose upon [man] any necessity to develop himself."[61] This brief comment seems to be the only occasion on which Marx raised the question why humid-tropical regions, including Africa, did not develop as Europe did.

The bountiful-tropics theory is still used today in some theories that try to explain the uniqueness of Europe's rise. I suppose that every European child of our own time has seen some version of the cartoon showing the native sitting under the coconut tree, waiting patiently for his food to drop into his hands. This is not merely a relic of oldtime thought or an implicit theory. Eric L. Jones uses the bountiful-tropics theory in *The European Miracle*. (In West Africa "living was easy."[62]) John A. Hall uses it.[63] Occasionally it is used by Marxists, faithful to the letter of Marx's comment quoted above.[64] Probably there is no need to explain why these bountiful-tropics theories are unacceptable, since this—as we have seen—is done time and again by scholars who insist that tropical regions are *not* bountiful; rather, they are barren and nasty. Neither the one view nor the other makes sense. And note that *both* are used toward the same end: to show that tropical regions have inferior potential in history.

Finally, we come to the theory that tropical environments are so disease-ridden that historical progress there is slowed, stopped, or prevented. Some historians develop this argument specifically with reference to Africa. Some others apply it sweepingly to all tropical regions: for Eric L. Jones it is a major reason why both Asia and Africa remained backward by comparison to Europe. Like the tropical-nastiness theories discussed previously, this one is traditional in European thought, and this fact is critically important for an understanding of the survival of the theory down to the present and its use in the European miracle paradigm.

One of the axioms of classical diffusionism, as we have seen in Chapter 1, is the idea that diseases and other evil things *naturally* counterdiffuse into Europe from non-Europe. It was therefore assumed by many scholars (including, for instance, Buckle) that non-Europe is both the source and the natural home of many—and the worst—maladies.[65] That axiomatic belief is still with us: plagues from the Black Death to AIDS are still assumed—for it is always an assumption, whether or not reinforced by evidence—to come from the non-European world.[66]

This foundation belief was reinforced during the eighteenth and nineteenth centuries by the reports of upper-class European travelers about the, to them, dirty, disgusting, diseased communities they found outside of Europe. This was both a class phenomenon and a cultural one: alien lifeways necessarily seemed to be unhealthy. But the belief gained powerful reinforcement during the nineteenth century, when the growing wealth and modernization of (much of) Europe led to a dramatic improvement in health conditions, grounded in general improvement in living conditions, improved sanitation, and, finally, progress in medicine. Europe, therefore, seemed to be somehow healthy, non-Europe, somehow unhealthy. This error is still very widespread. Underdeveloped countries are poor. With poverty comes ill health. But the ill health is thought, incorrectly, to somehow stem from the natural environments of these regions or from the cultures of their inhabitants, not from poverty. However disease-ridden India, Africa, China, etc., may have been in this period, Europe itself had been just as disease-ridden a century or so earlier (a fact known in part from the demographic facts about life expectancy of ordinary Europeans down to the eighteenth century).

In this same period European colonial territories were expanding, and it became evident that Europeans living in colonies tended to fall prey to various kinds of exotic diseases. The most extreme case apparently was West Africa, which was called in those times "the white man's grave." Here the insecurity of the quite small coastal settlements—colonial territories in sub-Saharan Africa did not really expand until very late in the nineteenth century—combined with the massive effects of the movement of slaves (many of whom died in these settlements) produced peculiarly unhealthy conditions, though the Europeans thought the source of the problem was the innate unhealthfulness of Africa itself.

Most of the important diseases of humans and their domesticated animals are *not* peculiarly tropical. Smallpox, typhoid, pneumonia, diphtheria, measles, bubonic plague, anthrax, and many other diseases are found across many physical environments, and their severity from place to place tends to reflect, more than anything else, conditions of human poverty, crowding, and the like. To some extent this is true even of the supposedly "tropical diseases," such as malaria (which, in fact, used to plague extratropical areas, including New York). Some forms of malaria tend to be associated with stagnant water, and thus with irrigated agriculture, in a wide range of climates. Other forms, including some of the most serious ones, are associated with tropical forest conditions, the mosquito vector often breeding in bromeliad growths on the trees themselves. It is true that these forms of malaria are especially associated with shifting cultivation in the humid tropics. But farmers spend relatively little time in the forest, if indeed there is forest. And they develop immunities such that malaria does not ravage their communities (as it does ravage communities of foreigners in their midst: traders, colonial military personnel, and so on).

The question is: after all such matters have been taken into consideration, is there, then, a remainder that can be called "the innate unhealthiness of the tropics?" Probably the answer is no.

Although this generalization is now widely accepted, some historians cling tenaciously to the idea that Africa is and always has been a uniquely disease-ridden place. This is given as one reason for Africa's (supposed) marginal role in world history, at least in modern world history. (According to William McNeill, disease "more than anything else, is why Africa remained backward in the development of civilization when compared to temperate lands."[67]) In the view of some of the absolutionist historians, notably Curtin, the disease-ridden character of West and Central Africa in and after the sixteenth century is a key part of the explanation for the complementary facts that this region became the source of slaves for the Atlantic plantation economy and that this region, instead of "rising" with the growth of the Atlantic economy, instead remained undeveloped. It is important to put this theory into perspective in terms of the confrontation mentioned at the beginning of this chapter between traditional European historians and the Third-Worldist, revisionist school. Traditional scholars (at least those of the absolutionist school) claim that Africa in the sixteenth century had a rather small and mobile population, with little development of complex civilization and state organization, and with slave raiding and slave trading as important features of its society. Africa had not risen above this very low level of civilization mainly because of the prevalence of human and animal diseases—with other environmental factors of the sort previously discussed being added on as additional factors. Africa thus quite naturally became the prime source of plantation slaves; and, also naturally, the slave trade did not fundamentally alter the reality of African life.

Third World historians tend to dispute all of this. Africa was densely populated before the slave trade, and the slave trade utterly devastated the continent, destroying states and civilizations, depopulating vast regions, and leading, overall, to disastrous underdevelopment. The slave trade mainly reflected Europe's power relative to coastal African societies in the seventeenth century and thereafter, in consequence of the immensely profitable American plantations and—as we discuss in Chapter 4—the fact that militarily and commercially powerful West African states were mostly located some distance from the coast. Disease *increased* in intensity because of the devastation caused by the slave trade: depopulation, the abandonment of large areas formerly cultivated and grazed (now turned to forest and scrub), wars, economic decline, and so on. One crucial example of this devastation has to do with the tsetse fly and trypanosomiasis or African sleeping sickness. Traditional scholars claim that tsetse fly infestation prevented Africans from developing extensive cattle herding in earlier times and in many other ways contributed to historical stagnation. The response is that trypanosomiasis, like anthrax, is a disease with many mammalian hosts. Depopulation led to the massive expansion of wastelands, and of wild animal host populations, and thus to the spread of conditions in which the tsetse fly could flourish. This in turn changed trypanosomiasis from an endemic disease to which both humans and cattle had some immunity and exposure, which was kept in check by the relatively full occupation of lands, into a devastating disease that, since the end of the last century, has indeed prevented the development of animal husbandry in some areas of Africa. There is considerable evidence that the expansion of bushland led to the expansion of tsetse-fly-infested areas and thus to economic and social misery.[68] Beyond this, there is little solid evidence either way about the history of health conditions in Africa and among Africans (not colonial visitors), and there are many reasons to doubt the inherited diffusionist assumptions and prejudices on this matter. In any event, one cannot make a case that disease was an independent force that "blocked" development in sub-Saharan Africa.

Arid, Despotic Asia

Asia is a large place and contains a large variety of environments. There is, understandably, a very long and varied list of traditional environmentalistic arguments, most of them particular to some one part of Asia and inapplicable to other parts. Back in the early nineteenth century, a pious form of environmental determinism prevailed in geography, and it seemed sensible to invoke a single explanation for all of this variety: God had placed different natural obstacles in the paths of different Asian peoples—heat in one place, cold in another, drought in a third. Today it is more common to find, among historians writing about the "European miracle," not a list of Asia's environmental infirmities but rather a set of separate comparative judgments, each centered on Europe, and each referring to a specific period in European history. Europe, or some part of Europe, at some particular time, was superior to all of Asia in environmental qualities X and Y and Z. Or, more typically, the comparison will invoke environmental obstacles for one Asian region, political obstacles for another, religious obstacles for a third, and so on, in a very eclectic sort of argument.

For these reasons I will not review all the environmentalistic theories as to why Asia supposedly remained backward in comparison to Europe. I will deal with the comparative judgments about climate, landforms, and so on, one by one, in the following section of this chapter ("Temperate Europe").

There is, however, one very large and coherent theory that is used today, much as it was in the last century, to deal with Asia as a whole in one grand and sweeping judgment of inferiority. This theory has a number of variants which are known by various names, among them "the Asiatic mode of production," "hydraulic society," and "Oriental despotism." The theory is not usually thought of as an example of environmental determinism, since it seems to start its argument with technology, claiming that irrigation-based ("hydraulic") societies have certain highly distinctive characteristics that inhibit historical development. But one of the roots of this argument is environmental. It is the claim that aridity in Asia made irrigation *necessary*. One can properly ask how a theory of this sort can be applied to parts of Asia that are not at all arid. (It has even been invoked as an explanation for Stalinism in wet, cold Russia.[69]) To understand this contradiction, and to understand accordingly why this theory is untenable in all its forms and variants, we must look briefly at the history of the doctrine.

European writers of the past half-millennium have tended to view Asia as a place where people are inherently unfree and society is inherently unchanging. It would take us far afield to go into the evolution of this belief, but by the eighteenth century it had become a significant part of the emerging doctrine of diffusionism.[70] It was accepted as an axiomatic truth, rarely questioned, but efforts were made to explain this inherent "Oriental despotism" (as it came to be called) in terms of everything from theology to race to environment. The belief seems to have been applied mainly to the Ottoman Empire, which was in that period a political and military threat to some European societies and a commercial threat to others, until the late eighteenth-century expansion of direct European colonialism in India and Southeast Asia gave the theory new functions. Not only was the notion of Oriental despotism useful as a justification and rationalization for colonial expansion, but it became the basis for colonial legal doctrines that were being fashioned at that time. In Asia, it was decided, there is no private property in land because the ruler, despotically, owns everything. Therefore, when we Europeans depose the ruler, *we* own everything. And if we take over a despotic state, we acquire the rights of despotic rule over a people who were unfree to begin with. (But European rule, however despotic— colonies had no democracy—was described as bestowing "freedom.")

The modern form of Oriental despotism was often connected back to the biblical "Orient." When modern Asian societies were described as "stagnant," it seemed fair to assert that they basically retained the character described for them in the Old Testament. (The Old Testament, we recall, spoke of the existence of great cities, empires, agriculture, etc.) While it was necessary to explain modern wonders like the Taj Mahal and great modern Asian states, it was not difficult to view these as relatively minor advances over the original biblical civilization, and to find secondary explanations for the evident fact that Asian civilizations had advanced *somewhat*—but long ago they had *ceased* to advance, and so remained essentially biblical and essentially stagnant.[71]

I think it likely that the geographical connection between aridity and "Orient" comes partly from this source, that is, from the biblical images of arid regions like Mesopotamia and Egypt. Partly, no doubt, it comes from the fact that early modern Europe, down through the mid-eighteenth century, thought mainly of the dry western Asian regions, the Ottoman realm and the Persian-Inner-Asian region, as "the Orient," because the farther Orient, from India to Japan, was still somewhat remote from European attention. In any event, early nineteenth-century geographers like the great Karl Ritter were describing one special type of geographical–cultural system, the type associated with Asian civilizations of the great river valleys of arid Asia and northeastern Africa, notably the Nile, the Tigris–Euphrates, the Indus, and smaller valleys of similar character, and ascribing the traditional Asiatic traits of despotism to regions of this type. They seem to have extended this model somehow to comprehend the river valleys of wetter parts of Asia, through a logical leap from the idea of irrigated river–valley civilizations in arid regions to irrigated river–valley civilizations in Asia as a whole.

The general association of Oriental despotism with river–valley civilizations, from Egypt to China, was commonplace in the nineteenth century.[72] Marx and Engels, however, took the idea a definitive step forward by advancing a theory that derived Oriental despotism from propositions about aridity and irrigation.[73] In the 1850s Marx and Engels were, for the first time, seriously confronting the question of how their essential theory of historical evolution could be applied on a world scale. It needs to be said first that they were probably the most skeptical European thinkers of their time as regards all of the traditional and (as they insisted) elitist social theories then in vogue. But their skepticism had inevitable limits, since they were products of an elitist German education and since they knew almost nothing about the world outside of Europe apart from what they learned in the press, and in books and official papers presenting the colonial point of view with all its prejudices. Accordingly, Marx and Engels did not seriously question the prevailing doctrine that the Orient is in some sense despotic and to some degree historically stagnant and unprogressive. But their skepticism about European social theory, with its elitist foundations, immunized them from the usual explanations for Asian despotism and stagnation. Asians were no less rational than Europeans, and no less willing to struggle against economic exploitation. This reasoning led Marx and Engels to speculate that the cause of Asiatic despotism and unprogressiveness lay, not in human society, but in the natural environment. In Asia, it appeared, social evolution had not led to private property ownership: the ruler, despotically, owned the land, except where it remained as original communal property. Hence this speculation from Engels:

> The absence of property in land is indeed the key to the whole of the East. Herein lies its political and religious history. But how does it come about that the Orientals did not arrive at landed property, even in its feudal form? I think it is mainly due to the climate, taken in connection with the nature of the soil, especially with the great stretches of desert which extend from the Sahara straight across Arabia, Persia, India and Tartary up to the highest Asian plateau. Artificial irrigation is here the first condition of agriculture. . . . An Oriental government never had more than three departments: finance (plunder at home), war (plunder at home and abroad), and public works.[74]

This theory was advanced very tentatively by Marx and Engels, and was modified in later writings; it appears that Engels rejected it altogether in his late writings.[75] Our interest lies in the fact that the environmentalistic component remains influential today, even though it is now obvious that Marx and Engels were mistaken in the notion that Asia is arid. The theory nonetheless has had an effect on recent Marxist discussions about the European rise of capitalism and the (supposed) nonrise of capitalism in Asia.[76] More crucially, it has been woven into the mainstream European miracle literature in an interesting intellectual sea change.

Max Weber, early in the present century, and Karl Wittfogel, at midcentury, are probably the key figures in the transformation of this rather archaic doctrine into a modern environmentalistic argument for the European miracle. Weber had little to say about the natural environment per se. Drawing on various scholarly ideas which were in circulation in turn-of-the-century Europe, including Marxian ideas, he argued that the development of private property in Europe's Antiquity and Middle Ages was indeed one of the primary features of social evolution toward capitalism. He stressed what he saw as a fundamental difference between the rise of feudal (seigneurial) property, which was close to, and moving toward, full private property, and a contrasting form that he associated with the Asian river–valley civilizations (and ancient Egypt). He saw that social form as being associated closely with the *need* to irrigate in such environments, hence associated, inferentially, with the environment. These latter societies were, he said, despotic and land was not, in general, passed fully into the hands of officials, but rather lent to them on temporary tenure, on condition of service, as a means of providing them with rental income, men for military levies, and so on.

Irrigation required collective labor appropriation under despotic rule, for maintenance of canals and waterworks. In forested lands of Europe, such despotic rule over peasants had not been needed.[77] Thus in Greece and elsewhere in Europe yeoman farmers, individualists, became the signature of rural society. Cities, instead of being mainly the seat of despotic power, became truly urban. Thus, in general, there was a peculiarly western trajectory, toward modern urban and capitalist society.

> The crucial factor which made Near Eastern development so different [from Greek development] was the need for irrigation systems, as a result of which the cities were closely connected with building canals and constant regulation of waters and rivers, all of which demanded the existence of a unified bureaucracy. There was an irreversible character to this development, and with it went subjugation of the individual. . . . On the other hand in Greece . . . the position of the monarchs declined . . . and so began a development which ended . . . with an army recruited from yeoman farmers who provided their own arms. Political power necessarily passed to this class, and therewith started to emerge that purely secular civilization which characterized Greek society and caused capitalist development in Greece to differ from that in the Near East.[78]

Weber's formulation of the difference between Oriental societies and Western societies has become one of the fundamental arguments of the miracle theory as it is put forward in our own time. But Weber did not put firmly in place one important part of the edifice. This is the matter of showing why irrigation societies acquired the special characteristics assigned to them: despotism and stagnation, and, beyond that, lack of private property, lack of full urban development, and so on. This essential technical and environmental elaboration of the theory was introduced in 1955 by Karl Wittfogel, in his book *Oriental Despotism*.

Wittfogel, an ex-Marxist who started his argument with the Marxian proposition which we have discussed, tried to show that societies grounded in irrigation, "hydraulic societies" as he called them, are necessarily despotic and must necessarily have the kinds of social and political properties that earlier writers had associated with Oriental despotism. The same is true of societies that somehow acquire-at-a-distance the characteristics of hydraulic societies, by diffusion. (Hence, according to Wittfogel, nonhydraulic Soviet Russia became despotic.) Wittfogel's reasoning was environmentalistic, in a manner less ignorant of physical geography than Marx and Weber but nonetheless quite naive. Wittfogel believed that irrigating a tract of land necessarily, deterministically, increases its productivity. Some societies will thus choose to adopt irrigated agriculture, and so the great river valleys of Asia were occupied. But, according to Wittfogel, irrigation requires major public works for the creation and maintenance of canals, and hence requires a command-type political structure—he says this was the origin of the state as a political form—which also functions to control the distribution of water.

Thus hydraulic societies are necessarily despotic. There are a number of fallacies in this argument, as many scholars have pointed out. Three of the fallacies concern the environmental proposition, the idea that irrigation necessarily increases productivity massively, and therefore leads to the development of complex social stratification, the state, and so on. First fallacy: irrigation increases productivity substantially only when the ecologically limiting factor on crop growth is lack of water; but quite often—in Asia as elsewhere—this is not the case. This means that the elaboration of great irrigation systems is not a natural response to the environment. Rather, a small-scale irrigation system in a river–valley can become enlarged into the great system as an *effect*, not a cause, of political and social inequality: pressure to deliver surplus pushes the process, enlarges the irrigation network, and leads, logically, to ever greater inequality. In other words, the large irrigation systems are preceded by, and explained by, existing despotism or social complexity, not by environmental mandates. (Wittfogel argues in roughly the opposite direction: the need to irrigate in a dry region leads a society to develop coercive command structures to manage the irrigation system, and thence leads to class oppression and the state—and despotic Oriental civilization.) The one ecological generalization that makes undoubted sense is the correlation of land productivity with population size and density, which is basic to the elaboration of social hierarchies and such things as religious ceremonial centers and states. But irrigation does not, magically, make land productive. Some land is highly productive without it. In the classical Mesopotamian and Nile cases we do not know whether the ruling classes forced increases in productivity, and this created the large irrigation systems, or whether the process worked the other way around. The second fallacy is the belief that the irrigation of truly dry river valleys made these valleys immensely productive, overall. Irrigation allowed agriculture to be practiced in lands otherwise desertic, but water was always in short supply and we know very little about crop yields. What we do know is that the really high productivity, and large social entities (in terms of population size as well as density), were found in wetter regions, mainly rice-farming regions, where elaborate irrigation systems usually were unnecessary for agriculture (and sometimes rainwater alone filled the paddies, as on the lower Irrawaddy plain and part of northern Luzon). Third fallacy: There is good reason to believe that some of the oldest civilizations were grounded not in irrigation, but in drainage, an ecological response that does not ordinarily involve large-scale water-works. Drainage systems seem to have preceded irrigation in the earliest Mesoamerican civilizations. It is quite possible that the earliest Eastern Hemisphere irrigation systems, in such places as the Nile, the Tigris–Euphrates, the Indus, the Wei, the Yellow, the Niger, etc., were originally drainage systems, opening up swampy riverine regions and perhaps elaborating polities long before the societies became committed to major engineering works and so became "hydraulic."[79]

Wittfogel's environmentalistic argument, bringing down to our own time the old ideas of Marx and Engels about aridity, irrigation, despotism, and stagnation, juxtaposed with Weber's arguments about the difference between Oriental despotism and the rest, is seminal for many present-day European miracle historians, although most of them modify Wittfogel's theory in important ways. The essential concept is the notion of hydraulic society, explicitly or implicitly seen as a natural product of arid Asian river valleys (and the Nile). The way this leads to "miracle" arguments in the writings of such historians as Eric L. Jones, Michael Mann, and John Hall deserves a moment's attention.

Eric L. Jones, in *The European Miracle*, lays stress on what he conceives to be the fundamental difference between "rainfall-farming" societies of Europe and irrigating societies of Asia. Wittfogel, he says, was essentially right. Irrigating societies suffered the "political consequences of a society with a huge, manipulated peasant mass."

> European agricultural society was able to avoid a comparable history of authoritarianism—a kind of political infantilism—by virtue of an open-ended productive environment of forest land and rainfall farming.[80]

Jones then adds to all of this his theory that farmers who work in warm standing water become diseased—a fallacy, as we saw above, that confuses the consequences of poverty with those of ecological settings.

The fairy tale (for it is that) which associates marvelous social consequences with "rainfall farming" of the European variety will claim our attention shortly. Here it should simply be noted that all of the supposedly dire effects that Jones—like Wittfogel before him—attributes to irrigation societies are, in point of fact, the normal attributes of ancient class society and ancient civilization. That is, when these civilizations emerge, we find, as part of the process, reduction in peasant freedom, recruiting of masses of people for various purposes, and the like. The romantic image of free peasants is an image drawn from preserfdom, prefeudal times. After Franco-Roman colonization European peasants were as unfree as any peasants of the Asian river valleys or anywhere else.

And of course rainfall farming, along with forested frontiers, was characteristic of much of Asia.

John A. Hall, in *Powers and Liberties: The Causes and Consequences of the Rise of the West*, wants to distance himself from Wittfogel's "fantasy" (as he calls it) about the inherently despotic nature of Oriental society. He then promptly absorbs a good part of Wittfogel's theory into his own formulation of reasons for the European "miracle." Hall, like most miracle theorists, tries to make use of the greatest possible range of traditional arguments for the miracle, then pushes certain arguments forward as supposedly the most important ones. He thinks that political forces and Malthusian demographic forces are the most important, although he sees as a deeper force the Weberian notion of European "rationality." Hall does not accept the formula that leads to the generalization that Oriental states were despotic. No, he says, they were arbitrary, cruel, and unwilling or unable to encourage economic development, and they held Oriental society stagnant (or, much the same thing, going through repetitive cycles). But they were not despotic. By this he means that Oriental states were not truly strong, appearances to the contrary notwithstanding. We will see, shortly, how he says that medieval European states had some sort of inner strength, an organic quality, with progressive change somehow teleologically prefigured in their medieval form. So, according to Hall, there was Oriental despotism but the despots were *weak*.[81] Hall then introduces irrigation as another independent factor.

> [In Europe] there was no need for irrigation. It is quite likely that this encouraged, or at least allowed for, a decentred agricultural civilization based on individual initiative."[82]

Thus a "need" for irrigation in Asia, which must have been an environmental need. Thus, the old Wittfogelian equation of irrigation with despotism (as against "individual initiative"). Then Hall moves to other factors on his very long list.

Michael Mann is another contemporary theorist of the "European miracle," and he too has a long laundry list of factors which, he thinks, contributed to the miracle.[83] He tends to emphasize the importance of ancient Europe's acquisition of political and especially military power. Mann takes pains to distance himself from Wittfogel, but in the end he incorporates most of Wittfogel's model—perhaps I should call it the Marx–Weber–Wittfogel model—into his own theory. Like Marx, Weber, Wittfogel, Jones, and Hall, he accepts the characterization of ancient Oriental societies (from Egypt to China) as unfree and unprogressive. But he points out that the despotisms did not extend to true large-scale politicomilitary power. And the ancient Oriental civilizations, although they were rooted in irrigation agriculture, also made use of other forms of resource use in areas adjoining the river valleys. Mainly for these reasons, Mann claims that Wittfogel "overextended his model."[84] Ancient irrigating societies were despotic but they were not powerful, and this, for Mann, is what counts. Mann then rephrases the model and puts it, more or less entire, into his own theory. The distinction that counts is indeed, he says, the one between "irrigating" and "rainfall" farming societies. Asian societies are sweepingly categorized as the former, European societies as the latter. According to Mann, the fact that European farmers, starting with the ancient Greeks, used iron plows and farmed unirrigated ("rainfall") land, is the one most fundamental reasons that Europe forged ahead of Asia and North Africa. This was the first great miracle. It put Europe ahead of all other areas, and Europe has remained ahead ever since.

Here, in brief, is Mann's argument. We start with the ancient Near Eastern civilizations, grounded mainly in irrigation. Conceding that irrigation was the innovation that caused these civilizations to rise in the first place, Mann argues that the irrigation base somehow "caged," or confined, the population; this metaphor is meant to convey the idea that these people are unfree, and also in some sense constrained from further social progress. (It is a metaphor and not an argument.) Around 1800 B.C., says Mann, "the Middle Eastern empires of domination were shaken by two immense challenges . . . from the north," from Indo-European invaders—this seems to be a version of the discredited "Aryan migrations" theory—who brought two revolutions with them, based in charioteer warfare and the use of iron tools and particularly iron plows. The "balance of power now shifted northward."[85] Mann concedes in passing that these northern folk did not actually *invent* chariot warfare and ironworking (which came perhaps from Anatolia), but he passes without pause (or logic) to the thesis that the northern peoples, Indo-Europeans, acquired dominance in politicomilitary terms and in terms of productive power. From this point forward, Mann contrasts two civilizations, the "irrigating" ones of the Middle East and the iron-plow–using, rainwater-farming peoples of the north—Greece and, broadly, Europe. Mainly because the Europeans were iron-plow–rainfall-farming peoples, they acquired, initially in Greece, modern civilization, including democracy, classes, private (more or less) property, science, and a respect for human reason.[86] Why? The centerpiece of the model is the image of an individual peasant-farming family which is fundamentally *independent*. It gets its water from the skies, not from a despotically managed irrigation system. Iron, says Mann, is abundant, so the peasant farmer does not depend upon cities and long-distance trade networks to acquire iron for plows and axes. This independent peasant, a true yeoman farmer (as Weber also had said), is the prototypical democratic, civilized, energetic, forward-looking *European*.

None of this makes sense. To begin with, it is geographically absurd to imagine that the Middle East was a region without plow agriculture, one basically of irrigated river–valley populations. Plows were used in irrigated farming. Rainfall-based farming was dominant, not in Egypt and Mesopotamia, but in most of the Levant, Anatolia, Iran, and of course much of Asia farther east, not to mention Africa. Iron working was not invented by Europeans and was used as much by non-Europeans as by Europeans. The same is true of the iron plow, which was as important in early China, for instance, as in early Europe.[87] (It is an old myth that Indo-European speakers, Aryans, spread the use of the plow to, and only to, the regions they settled or conquered.) Workable-grade iron ore deposits are not as abundant as Mann thinks they are (except in certain special regions, such as some of the laterite deposits of the humid tropics).

But the silliest part of Mann's thesis is the environmental determinism. We have an image of an arid region fit for nothing but irrigated farming, which is assumed to be a tether on the progress of civilization. We have an image of an open region of good soil, forested, in which iron-using peasant farmers not only produce an unmatched abundance but also acquire from their ecology a democratic, bold, bright, form of society which marches forward then to modernity. Let me add just a word about each of these images.

Irrigation, in the eyes of many miracle theorists, is somehow unproductive. Jones, Mann, and others make the assertion that more food is produced per worker on rain-watered lands. This is not true. In fact, irrigation was developed to increase food production, whether or not the source of the process was the aspirations of village people themselves or the surplus requirements of a political or religious superstructure. When an irrigated district has reached the point where there is serious land shortage, that is, late in its social and geographic evolution (along one possible line of development), the productivity per person declines. This is obvious. It means, first of all, that farm workers spend more time during the year producing and have less time for nonfarming activities, including village cultural life and work on such things as monumental structures. Eventually this situation may lead to the collapse of the farming system, with salinization, and so on, and finally, perhaps, starvation. Contrast this now with nonirrigated agriculture. Normally, there is poorer nutrient status of soil (irrigation brings dissolved nutrients in the water, and alluvial soils tend to have good nutrient status). Normally, there is greater moisture deficit, given the dependence on unpredictable rains. Thus natural fertility tends to be lower on nonirrigated land, all other things being equal (which they often are not). Now it is true that early iron-age farmers, clearing forested land and making new farms, achieved high production on this fresh land, but that phenomenon is transitory, and in any case it did not have much effect in the semi-arid valleys of Greece. Early European farmers tended in fact to settle on alluvial lands and lowland terraces, and quite early they were using irrigation and drainage—because it enhanced labor productivity.

The notion that European agriculture was somehow conducive to independent living in a way contrasted with Asian agriculture is another part of this myth. Early European farmers were not to any great extent, as Mann depicts them, living in isolation from one another, surrounded by forest. In the cases where they did live this way it was, again, a frontier phenomenon, in Asia as well as Europe. Farmers mostly lived in social aggregates, villages, large or small, compact or linear, depending upon many circumstances. There is no reason to doubt that the upland farming communities of Asia were quite similar to those in Europe. We return to this matter later in the chapter when we discuss the myth of the unique European family. For now I will simply categorize as mythic the idea that early European agricultural communities were somehow more individualistic, more independent-minded, more progressive than were communities elsewhere.

The most fundamental error made by the nineteenth-century thinkers and Weber and now mechanically repeated by historians such as Jones, Hall, and Mann, is to believe, or assume, that one type of environment produces a particular type of society and the latter then persists down through history. You simply cannot contrast the very ancient irrigating civilizations of Asia and North Africa with the *later* nonirrigating farming civilizations of Europe—or indeed of Asia—and then suppose that two contrasting civilizational types have been, thereby, created so as to remain in place down through history. Culture changes. Farmers move from one environment to another. In many places farmers practice *both* irrigation and nonirrigated agriculture on different soil types when appropriate land is available to them.[88] Thus the theory which asserts that arid Asian agricultural civilization produces a stagnant, despotic form of society down through later times, one which will not develop toward modernity, is purely a myth.

Temperate Europe

We have already seen how historians constructed a mythic model of "rainfall-farming" European society, supposing that rainfall brings benefits not associated with irrigation and that Europeans alone, in their hemisphere, practice "rainfall farming." Historians like Mann, Jones, and

Hall carry the argument farther along. Europe's rainfall-based farming, combined with Europe's supposedly fine and fertile soils, produced an environmental basis for agricultural production unmatched elsewhere. Although these historians give a lot of credit to the supposedly uniquely rational, inventive, European mind, they give much of the credit to Europe's natural environment.

For example, Michael Mann draws a picture of a steady northwestward movement of European history—what we described in Chapter 1 as an Orient Express model—which he sees as an essential continuance down into the Middle Ages of the northwestward trajectory of that individualistic, creative, bumptious peasant society which, he thinks, emerged in the Iron Age with the uniquely European marriage of farming, iron, and rain-watered land. His model of the evolution of Europe's society (its "miracle") is much more complex than this, but a key part of that model is the inexorable, steady, historically pregnant geographical movement—what he calls a northwestward "drift," with a permanent northwest "leading edge." Eschewing philosophical determinism, he nonetheless gives the whole process a strongly Hegelian, teleological flavor.[89] One of two main reasons for this northwestward movement is the beckoningly fine environment of northwest Europe. And the main reason why this environment is so fine is its "deeper, wetter, more fertile soils."[90]

John Hall, similarly, extols the "northern European clay soils,"[91] northwest Europe's "deep and productive clay soils fed by rainfall" ("There was no need for irrigation").[92] Eric L. Jones, in *The European Miracle*, makes much the same claim, if slightly qualified. He writes of Europe's "open-ended productive environment of forest land and rainfall farming,"[93] of Europe's "High, even rainfall and passable summers."[94] Jones (unlike Mann and Hall) recognizes the fact that Asian land supports a higher farming population, with the implication that the land is more productive, but he merely notes that Europe's productive land comes in smaller, separated regions, and then he gives us one of the myths of Oriental despotism:

> The very impracticability of hydraulic agriculture freed a fraction of European energies for other purposes. The rainfall farmers of Europe might be fewer in number than the farmers of China and India, but the former spent less time on all aspects of farmwork than the latter spent on water control work.[95]

The implication, of course, is that European peasant farmers didn't have to spend much time at farm work in order to satisfy their needs and their quota of surplus, whereas Asians worked much harder for the same product. This is plainly absurd unless we abandon historical method and try to compare Asian peasants under modern conditions of extreme land shortage with ancient European frontier conditions or modern European capitalist farmers. Stated differently, this makes no sense in terms of actual agricultural technique—irrigation is intended to increase productivity per worker, and usually does so—and in terms of any model of the historical European peasant as poor, oppressed, and overburdened.

The "wet soils" of which these historians speak are usually excessively wet: acidic podzols and gleys that are difficult to work and infertile until heavily manured. The "high, even rainfall and passable summers" are in fact a climate so wet that solar energy is often severely limited, grain crops sometimes cannot do well (recall how important was the post-1492 introduction of the potato), and soils do not dry out until late in spring, if at all. I do not want to caricature the situation. It is merely the case that northwest European soils, across their entire range of variability, are not superior to the soils of many other regions. In ecological terms, the lands of warmer, drier regions that have either good rainfall regimes or possibilities for irrigation or drainage tend to be rather higher in productivity. My point is more limited. There is nothing about Europe's agricultural environment that explains the so-called miracle, or that might lead one to believe (as the authors I have cited believe) that European history throughout its course has been favored over Asian history by the European agricultural environment.

Europe's supposed environmental superiority is by no means limited to farming. I will close our discussion of environmentalism by briefly commenting on four other comparative judgments which are routinely made by present historians of the European "miracle."

First is the classical "capes and bays" argument, familiar to most European and Anglo-American schoolchildren. Supposedly, Europe's configuration of peninsulas and bays, and Europe's possession of navigable rivers, gave this continent a natural basis for communication, trade, and accessibility denied to other continents. This, then, is supposed to have had a lot to do with the rise of markets and eventually capitalism in Europe.[96] Part of the argument, like the emperor's clothes, is transparently false when looked at directly. For instance, accessibility by sea along the coasts of the Indian Ocean and much of the South China Sea was considerably easier in the Middle Ages than it was in Atlantic Europe. (If monsoonal wind patterns posed a problem in the Indian Ocean, storms posed a greater problem in the North Atlantic.) Trading cities dotted the Indian coast, and trading vessels plied up and down this coast, and on to Indonesia and Arabia, carrying bulk commodities like rice and iron as well as high-value goods, long before Italian galleys inaugurated a regular commodity trade between the Mediterranean and the northwest

Atlantic. I do not belittle the great seamanship of the Atlantic fishermen, the Iceland traders, and the Hanseatic League when I note, simply, that other regions were using the sea as effectively as the Europeans were at comparable times.

As to rivers, Europe's endowment of navigable rivers is impressive but not unique. It is not better than India's and China's. Interisland navigation in insular Southeast Asia was much easier than was navigation up and down the Rhine, say, or the Danube. Again it is a matter simply of reducing comparisons to reality. The "capes and bays" errors are normally followed by another error: the idea that Europe had, in early times, an advantage over civilizations that had to transport commodities overland. Many of the "miracle" historians repeat the theory that overland transport necessarily was much more costly than water transport. This is in fact a multiplicity of comparisons. When Mann and others quote the old formulas about draft animals consuming their weight in fodder over a limited distance (some say this takes place within 100–150 km), they ignore the fact that most draft animals graze or browse along the way: hence the great Inner-Asian Silk Road and the Sudanic caravans are easily comprehensible.[97] Canals in China were much more effective than circuitous coastal shipping routes in Europe. And water transport—particularly upwind and upriver—may not have had advantages over land transport 1,000 or so years ago.

Eric L. Jones, in *The European Miracle*, makes the large claim that Asia suffered so much more than Europe from natural disasters of all sorts that Asian development was strongly inhibited by this presumed fact.[98] He carries the argument to the point of claiming that the risk of natural disaster was so high as to frighten ordinary people into unusual demographic behavior, to close off trade possibilities, and more. Setting aside the fact that we have very little historical data on this subject, the main error is a simple matter of scale. Jones's "Europe" is really West and Central Europe. This region is roughly the size of the Indian subcontinent and perhaps one-fourth the size of the settled portion of Asia. So it might well have one-fourth as many natural disasters, *ceteris paribus*. It is true that floods are more serious in regions where people farm river valleys, but populations make adjustments which keep this one risk small enough so that it does not affect long-term development, and I suspect that winter-weather dangers in rural Europe were as great as flood risk in Asia, on a per capita basis. Hurricanes are no worse than the worst North Atlantic winter storms. The theory as a whole is unsupported by evidence and empty of credibility as an argument for the European "miracle."

The miracle historians point repeatedly to Europe's environmental differentiation, claiming that this leads to a unique potential for interregional trade and thus, again, for capitalism.[99] One need merely note that Europe does *not* possess an unusually large range of environments, and natural products, by comparison with other regions of comparable size. China, for instance, has roughly the same range of midlatitude environments plus a tropical south coast.

The last environmentalistic fable to be discussed here is the claim that Europe's topographic differentiation into many small "core regions," separated by mountains and forests, somehow led to a number of supposedly unique features of ancient and medieval European society, including even the development of a unique trading system and a unique system of moderate-sized states and the many benefits supposedly deriving therefrom. Since this issue ties in with the parts of the "miracle" literature that concern Europe's supposed uniqueness in economic and political development, I will postpone discussing the geography of the matter until later in this chapter. Suffice it to say now that this idea of "cores" is partly a myth, and insofar as it is valid the same sorts of "cores" can be found in other continents.

Rationality

We turn now to the theories about Europe's historical superiority or priority—Europe's "miracle"—which ground themselves, not in biology, not in environment, but in culture. First I will deal with theories which start from a conception of the superior "rationality" of Europeans. "Rationality" in such theories embraces many psychological attributes, always including inventiveness and innovativeness (or progressiveness), usually a capacity for abstract thought, and often a certain ability to make moral or ethical judgments. The issue is not whether such things are or are not fundamental causes in history; the issue is whether Europeans had *more* rationality, or *higher* rationality, than every other human community, and whether this was the principal reason, or at any rate one of the principal reasons, for the unique rise of Europe. The plain assertion that Europeans are smarter than everyone else seems to have a certain antique or Victorian ring to it, but this is deceptive: theories about "Western rationality" and the like are just as important today as they were for prior generations of scholars. Sometimes they are difficult to identify as rationality theories because they display other plumage. An explanation may focus on, say, technology, arguing that European technological innovations produced various forward movements in European history; looked at closely, however, the technological explanation usually dissolves into a theory about the inventiveness of Europeans, that is, their rationality. By the same token, an explanation may start with the state, or free markets, or the family, but usually (not always) it derives such structures from the more basic trait, rationality. Sometimes it works the other way around. Classical racist theories were rationality theories: to be white is to be more intelligent. Some (not all) Marxist theories are rationality theories: the defeat of feudalism released creative energies which then led to technological innovation, etc. Another complication is that superior rationality may be posited just for one crucial period and place, a magic key that started the whole process: Periclean Athens, Gutenberg's workshop, etc. I will try to sort these matters out by dealing first with the category "rationality theories" as such, then turning to other sorts of theories, technological and institutional, some of which are also rooted in the idea of European rationality.

The Rationality Doctrine

At the beginning of this century most European scholars accepted the basic proposition that Europeans are more rational than non-Europeans. This was explained by a great variety of competing theories, biological racism being perhaps the most prominent, but I suspect that most Europeans accepted the proposition as indisputably correct regardless of the explanation. It was simply obvious that European countries had attained a higher level of wealth and civilization than other countries and had done so on their own, mainly through invention, innovation, and creativity. Europeans also now controlled the entire world and this, too, had to reflect some intellectual and probably also moral superiority. This was the heyday of the doctrine of classical diffusionism (discussed in Chapter 1), and few people doubted the doctrine's basic propositions: Europe develops; non-Europe does not develop or does so more slowly; Europe's development is based ultimately in some intellectual or spiritual principle; the normal and natural way for non-Europe to modernize and progress is by receiving the diffusion of rational European ideas, brought by European colonial administrators, settlers, planters, missionaries, and purveyors of commodities.

By this time most European thinkers had come to accept the doctrine of "the psychic unity of mankind," at least to the extent of agreeing that all of humanity shares a common ability to progress toward modernity. This doctrine crystallized into a theory, widely though not universally accepted in the early part of this century, that can be called the dualistic–developmental conception of human rationality. The elementary dualism was a distinction between the mentality of child and that of adult. The human mind has developed from a prehistoric condition which was mental childhood. European history is either to be explained as the fruit of human mental development or has been intimately accompanied by such mental development, in a process that was fundamentally the same as the psychological development from childhood to adulthood. Europeans became more rational as history progressed, just as children acquire rationality in the course of ontogenetic development. Ancient people had been not merely less intelligent but also much more governed by emotions and passions than by intellect, just as is the case with modern children. With some modification, the same was thought to be the case for modern European women, who were less intelligent and more governed by emotion than men, that is, they were less rational than men. But women, too, would experience mental development, and would eventually be rational enough to vote, hold public office, etc.

Non-Europeans, within the same theory, were seen as psychically *undeveloped*, as more or less *childlike*. But, given the psychic unity of mankind, non-Europeans could of course be brought to adulthood, to rationality, to modernity, through a set of learning experiences, mainly colonial. (The phrase "colonial tutelage" was a signature of the doctrine, and this conception is encountered in most history and geography textbooks of the time.) It was not simply a case of "the natives are like children." The idea of non-European nonrationality was a definite, putatively scientific principle, widely accepted: non-Europeans think somewhat like children, and will be led toward adulthood by Europeans. Non-Europeans were of course graded. "Savages" were mental children without qualification. Problematic peoples, like the Indians, Ottomans, and Chinese, were thought to be childlike in some respects and not in others. Indeed they were governed by emotion and passion much more than Europeans. (Colonial revolts were obviously irrational—were outbursts of childlike emotion.) As far as scientific and abstract philosophical thinking was concerned, the people of these cultures were clearly not up to full adult standards, but in some respects, notably in arts and crafts, they were perhaps gifted adolescents.

So we have a model which has, as its centerpiece, the Rational Modern Adult European Man. History is his progression to mental adulthood. He is contrasted with ancient European man; with modern European children; with modern European women; with modern non-Europeans. The contrast was often extended, also, to psychotics; in some schools of they thought were explicitly seen as having a developmentally arrested mentality and this was indeed used as a therapeutic principle.[100] What needs to be emphasized is the fact that this was considered to be a scientific theory. Thus it was valid to exchange principles across these various dimensions of contrast. For our purposes, the most important of these principles was the attribution of mental development to European history and the attribution of mental nondevelopment to non-Europe, past and present; hence Europeans had *naturally* acquired a unique rationality.

It would be wrong to reduce to caricature the scholarly theories of a few generations back, and the model I have just described looks, on the face of it, like a caricature. Most scholars did not literally believe, say, that the American Indian was a mental child, or that ontogenetic mental development exactly recapitulates the mental development of our species. But the model was a dominant force in European thought in certain precise ways. The equation of child, ancient, and non-European was explicitly accepted in nonscholarly discourse (newspapers, Tarzan novels, and so on). American educational policy toward Indians and other colonials was explicitly grounded in this model, even if it was a smoothed-down version of it. A very large number of writings in history, geography, and all of social science made use of one or another part of the model. And several very widely accepted theories were really forms of the same model. Three examples will be helpful.

Most anthropologists accepted the idea that there are two distinct mental types, one being the so-called "primitive mind," the mind of tribal peoples just about everywhere (some added: peasants everywhere). The "primitive mind" was very explicitly described—the most famous description is found in Lévy-Bruhl's influential book, *How Natives Think*—as being incapable of higher theoretical and abstract ideas, as being emotion driven, and so on.[101] Some anthropologists opposed this notion (Boas, Radin, and Mead perhaps most notably[102]); most anthropologists accepted it with some qualifications; some accepted it literally. Lévy-Bruhl's rather stark form of the theory had great influence on psychologists and all of social science. Closely connected to this theory was the notion that there are "primitive languages," languages incapable of expressing higher theoretical and abstract thought. This old notion (which had been used in one form by William von Humboldt) was joined to the proposition that people cannot think beyond the limitations of their natural language, and so a primitive language entails a primitive mind.[103] Beyond that, the old philological theory about the innate superiority of Indo-European languages still had many adherents, and this theory extended the notion of primitive language to most of the non-European languages of the world. The third example is a broad family of theories in psychology that made use of the idea that mental development in the individual is homologous to mental development in the species, and that modern primitive peoples have the mentality of children and ancients, sometimes extending the concept to the putative psychological makeup (and limitations) of non-European peoples in general. In the twentieth century this theory received its most influential expression in the psychoanalytic viewpoint of Carl Jung and his followers. Jung swept broadly across non-Western cultures (Arabs, Indians, Africans, African-Americans, and others) declaring, basically, that only modern European man has fully developed an individual consciousness, an ego, an ability to think, even an ability to conceive of himself as an individual separate from the external world. Only European man is rational.[104]

The doctrine went through some important changes as it became absorbed into the "modernization" paradigm, the body of ideas which, as we have seen, came to dominate European social thought in the 1950s and still does so to some extent today. "Colonial tutelage" gave way to "diffusion of modernizing innovations." Non-Europeans no longer were "natives," and no longer were described as "childlike." In place of the notions of "primitive mind" and "primitive language" came the notion of *traditional mentality*. Non-Europeans are "traditional" in two senses: they lack "modern cognitive abilities," that is, the ability to think theoretically and scientifically, and they lack "modern attitudes" of the sort that push a person to achieve higher things, to reject the old, and so on.

This appears still to be the primitive mind, but there is an important difference. "Traditional minds" are simply waiting to be awakened, to be modernized. The larger picture was one of a vast landscape of traditional societies, containing people with traditional mentalities, and modernization would set the drama into motion. Mentalities would change. Social structure would change. New ideas and technology would diffuse into the modernizing societies from the already-modernized European societies, and so on. Recall that the modernization doctrine is not only a concept of the spatial spread of European ideas, attitudes, and the like, in the present-day postcolonial world. It is also a historical concept: modernization as history. "Traditional society" and "traditional mentality" came to be used to describe early Europe, the rise of Europe was now seen as a modernization process, and the *start* of that process was a much debated matter of deciding when Europe began its "takeoff"—its rise from the level of traditionalism and toward modernity. History is appealed to for the basic causal principle: Europe modernized, and now non-Europe will follow—though not slavishly—in its path. We will come back to this point in a moment. First I want to comment on the main causal principles that are thought to lead in the other direction. I will show that schools have arisen in psychology, sociology, and other disciplines that claim to provide scientific demonstrations as to what, exactly, is the nature of the "traditional mind," and the output from these scholars tends, then, to seep back into history, where it finds its way into the newer writings about the European miracle.

In psychology, the important figure is Jean Piaget. His basic theory of mental development postulated an invariant series of "stages" of development through which all children must progress. Piaget was influenced by Lévy-Bruhl and the primitive mind doctrine, and as late as 1971 he thought that

> it is quite possible, and this is the impression we have from known ethnographical work, that in many societies, adult thought does not go beyond the level of "concrete" operations, and therefore does not reach that of propositional operations which develop between the ages of twelve and fifteen in our milieus.[105]

"Propositional operations" means, roughly, logic. "Concrete" is childish prelogical thought, unable to deal with the abstract and the theoretical. Now Piaget was a great psychologist, but he did not have, or claim to have, direct knowledge of non-European psychology. He was arguing from the traditional dualistic–developmental model that equates the child and the primitive. (And he was about 30 years out of date as to ethnography.) But many of his disciples, seeking to connect their theory with the modernization doctrine, took up the theme and began to carry out research to find out whether non-Western people (particularly Africans) do, indeed, have inferior cognitive abilities. It is no oversimplification to say that virtually all of these studies made the same mistake and came up with the same, predictable conclusions. They used the tests of cognitive ability that Piaget had used with European children and, scarcely modifying them, administered these tests to non-European children and adults and found, predictably, that these people do not have full adult cognitive abilities. By now the error has become well known, and most present-day Piagetian psychologists do not seem to assert that non-Europeans are cognitively childlike. Along with the Piagetian studies of the 1960s and 1970s, there appeared, in this era, many studies carried out by psychologists from other schools of thought (like the Heinz Werner school) which took the old primitive-mind doctrine as fact,[106] along with studies carried out by white South African psychologists comparing white and black subjects, other Europeans studying other African populations, and a few Israeli psychologists comparing the cognitive abilities of Arabs and Jews. With few exceptions, the European psychologists found the non-European subjects to be deficient in cognitive ability, and so to be "traditional."[107] In nearly all of these studies the testing was such that the "natives" didn't really stand a chance because the tests were European, administered by Europeans or their native assistants, in Europeanized settings. The error became so well known that the entire field of "cross-cultural psychology" gained a bad name—as an apologia for ethnocentrism—until, in the 1980s, it changed direction. The point is that while most cross-cultural psychologists today, apparently, deny that non-Europeans are less rational than Europeans, they established a body of publications that are still used by others to make the opposite point.

The rationality doctrine was perhaps most influential in those parts of sociology and economics that were closest to the modernization doctrine, particularly among academics who were involved in making and implementing the policies being put into place in the 1950s and 1960s to develop the underdeveloped countries. As we saw in Chapter 1, this sphere of ideas was crucial to the politics of modernization for at least three reasons: First and most basic was the need to validate diffusionism. Second, development of the sort that involved only the spread of new ideas and new techniques was much cheaper, in principle, than development involving massive flows of capital, industrial development, and the like. And third, along the same lines, development at the level of ideas, of research, extension, education, and so on, was not *threatening*; it would not produce the dangers of revolution and counterrevolution implicit in efforts to change the relation of power groups, to effectuate land reform, and the like. For these reasons, scholars were encouraged, indeed were given lavish grants and were well paid, to produce analyses of modernization that would lead to workable policies for bringing about development mainly through an influence at the level of ideas.

Everett Rogers, a rural sociologist; David McClelland, a social psychologist; and Everett Hagen, an economist, provided highly influential contributions that I will describe briefly and with some caricature. Rogers was one of the leaders in a movement to sort out peasant mentalities into those that are prodevelopment ("cosmopolitan") and those that are noninnovative and "laggard." The crucial notion was the idea of the diffusion of rationality into rural non-European communities. The key to development (with some qualification) was the transmittal of new ideas to innovative "adopters." The fact that most of the ideas were not, themselves, workable (thus were not rational), and that adoption of them would have required of peasants not more knowledge, but more power and landownership, was ignored. McClelland claimed that non-Western peoples in general have not modernized because they lack the proper "need-achievement" motivation. They will modernize when and if they acquire this need to achieve, which had been of crucial importance in earlier European development, from ancient Greece on up. Hagen produced an elaborate theory, based in no evidence whatever, that non-Western peoples, and particularly peasants, have a traditional mentality that resists change of all sorts. The model starts with a mythic peasant unconscious mind that is uncreative, submissive to paternal authority (and therefore oppressive of innovative children), and unwilling to change. All three of these scholars, and dozens of their acolytes, produced a general model of peasant mentality as defective in both cognitive and attitudinal (affective) characteristics, but, of course, not hopelessly so.[108] One of many recent outgrowths of this doctrine is Stephen Marglin's theory of the Western mental "episteme," a marriage of the old "primitive mind" idea and the newer ideas about traditionalism, which argues, in essence, that the Western mind is, and historically has been, scientific, rational, and intellectually hardheaded—the "episteme"—while the non-Western mind is traditionally given to technique and art ("techne") but not to science. Marglin's new twist on this old doctrine is to say that some of the non-Western "techne" should be incorporated in development along with the Western "episteme" because "techne," less intellectual but more feeling, is less likely to destroy the natural environment and causes less psychic damage to non-Western people, a distinction rather like the older duality between scientific thought and emotion-laden concrete thought.[109] Marglin's ideas provide an example of much theorizing in the economic development world (like Hagen, he is a development economist) about non-Western nonrationality.

It would take us far afield to examine the parallel arguments in all the other fields of social thought. The following should suffice: In geography, the diffusion-of-innovative-ideas approach has been important since the 1960s and remains so today. The same old assumptions about peasant and non-Western traditionalism are still dominant. Some geographers claim even that non-Europeans are mentally not capable of using all available means of coping with natural hazards like drought and hurricane. One geographer, Robert Sack, has come up with a classically Eurocentric theory about spatial cognition: most non-Western people (primitives and most peasants) cannot think in spatial terms the way modern Western adults can. Among the authorities he uses—the argument is from authority, not evidence—are Lévy-Bruhl and Piaget.[110] In the field of education in the United States the newer forms of the dualistic–developmental theory of (Western) rationality are highly influential in many areas, including testing.[111] In the field of philosophy, there is a strong relationship between the stream of ideas about rationality, discussed above, and the resurgence of mind–body dualism, such as one finds in the Cartesian and Kantian traditions and particularly in the modern neo-Kantians.[112]

I do not suggest that the doctrine of non-Western nonrationality is fully hegemonic in modern scholarship, or that the above formulations have not been challenged. In general, however, the doctrine remains somewhat dominant in all fields of thought except, perhaps, anthropology and economics. Anthropology in recent decades has, with some lapses, maintained quite firmly that primitive minds are *not* primitive.[113] And some schools of economics need the principle of universal economic rationality so badly, as axioms for their theories, that they are willing to concede rationality to everyone.

Rationality and the European Miracle

The idea of European rationality is often called "Weberian," because Max Weber made important use of the idea in various explanations for European social evolution and various negative judgments about the lesser rationality of other societies. The rationality doctrine was, however, widely held in early twentieth-century Europe when Weber's important writings on this subject were published. But he codified the traditional doctrine and added something of his own to it, so the doctrine can indeed be called Weberian from the present-day perspective. It is also Weberian in another and perhaps more important sense: Weber placed European social evolution in a framework emphasizing the modernization process, and contrasting Europe's modernization with the "traditionalism," as he sometimes called it, of Asian civilizations. When the modernization paradigm locked into place among scholars just after World War II, Max Weber was the most obvious, and most logical, source of basic sociological doctrine for the paradigm. From that time on, scholars who wrote about the rise of Europe and chose to emphasize the individual trait of rationality, or social-level traits and institutions which Weber had treated as primary (and, in the main, as consequences of European rationality), were Weberians to one degree or another (even if they rejected some parts of Weber's model). The importance of Weber for our discussion of present-day theories about the "European miracle," is the fact that the most influential, probably dominant, school of thought concerning the "miracle" today is more or less Weberian. Weber's views about rationality therefore require at least brief discussion.

Weber analyzed Western capitalist society of the late nineteenth- to early twentieth-century epoch in meticulous detail and with great insight, although there were important limits to his insight. He held to the typical conceit of his time and place and class in thinking that contemporary European capitalism is the culmination of a process of social evolution that was, at root, an intellectual progression, an ascent of human "rationality," meaning intellect and ethics, from ancient to modern society. (He had doubts about the likelihood of further ascent in the future.) At each stage of that progression people invented new social forms, such as higher forms of the state, the legal system, the bureaucracy, the economy, the city, and so forth, but these forms were, basically, products of evolving rationality rather than primary causes of progress in their own right. But the march toward ever more rational society took place in Europe, among Europeans. Outside of the European tunnel of time, all societies were in varying degrees traditional and in varying degrees irrational.

Weber said rather little about the question of why Europeans came to display this rationality in the first place, that is, as a basal cause. He invoked several factors, one of which, as we noted previously, was race and another, the natural environment. Others seem to lie deep within culture. Yet Weber rarely discussed causality at this basal level, that is, attempting to explain why Europeans have unique rationality, and have had it since a long time before the Protestant Reformation and modernity.[114] One reason seems to lie in his view of social causality, with ideas and values, and the evolution of ideas and values, treated as prime causes of social processes, social structures, and social change. Given this conception, he would not be expected to look for nonideological-level causes for ideas and values. Doubtless there were other reasons.

Whatever Weber considered to have been the basic cause of the differences between rational, progressive European society and irrational, traditional Asian society (Africa and America were scarcely noticed), he delineated these supposed differences very carefully. The most crucial arguments are found in his *General Economic History* and *The Protestant Ethic and the Spirit of Capitalism*. The development of rationality among Europeans—however that had happened—led to a special sort of "economic ethic," a body of values, aspirations, and logical thought processes that emerged primarily in connection with the Reformation (and particularly Puritanism) but which, more fundamentally, produced capitalism. The important point here is that basal rationality produced both the "economic ethic" of capitalism and the Protestant movement: Weber is not (as some think) explaining capitalism and modernity in terms narrowly of religion. He does invoke religion to explain many aspects of the supposed traditionalism of Asians, but here too a primordial irrationality is seen as underlying religion. (He writes, for instance, of the "magical traditionalism" of Indians and Chinese.[115])

The superior rationality of Europeans produces other major historical differences between Europe and non-Europe. Europeans are basically freedom loving, not ground under by Oriental despotism; this devolves into a form of city that is much freer than the Asian city, the latter being conceived by Weber as (in essence) merely a physical entity wholly controlled by the overarching empire, and the former as a truly new form of society, emerging in the Middle Ages and inaugurating modernity.[116] (As we discuss in Chapter 3, Asian cities were not as Weber described them. Some were free city–states, others were virtually free under a loose imperial umbrella. European cities, on the other hand, were much *less* independent than Weber thought.) Landholding systems were also different because of differences in rationality: only in Europe, says Weber (here putting forward a purely traditional European view) does the concept of private property truly emerge. The feudal estate is almost genuinely private property. The Asian estate is seen as merely a temporary assignment of land as a source of revenue for dignitaries, a form of salary given for service and ordinarily returned to the state. But European feudal estates were also, legally and actually, granted on service tenure, and in both continents these sorts of estates tended in general to become hereditary and thus eventually private property (see Chapter 3). Of course, Weber talks about other supposed differences between Europe and non-Europe, but the examples given here will suffice to show his overall approach.

Weber did not originate most of these arguments, although he codified them and developed them in a brilliant way. Yet Weber is given credit overall, and "Weberianism" is now the most important form of theory in explaining the European "miracle." The basic form can be summarized very simply. Rationality is the causal root. The effect of differential rationality is, in Europe, permanent progress, modernization, and capitalism. In non-Europe it is stagnation, traditionalism, and various irrational cultural attributes such as superstition. The model here is simple diffusionism: rationality and therefore modernization are permanent in Europe. Non-Europe does not modernize except by the diffusion of these traits from Europe.

The doctrine of "Western rationality" is widely used today in explanations for Europe's unique rise, its "miracle." No longer is the superior rationality of Europeans attributed, even implicitly, to racial superiority, but how the historians manage to assert the one without the other is not an easy question to answer. Generally, causality is consigned to the impenetrable mists of ancient history, with perhaps an occasional speculation about ancient free-living European peasants or the evils of Oriental despotism, or with a ritual citation of Max Weber. For many historians, I suspect, the idea of European rationality is simply axiomatic. Europeans, for whatever reason, are just built that way.

Eric L. Jones dwells very heavily on rationality as an explanation for

Europe's historical superiority in *The European Miracle*. His claim is that there is a fundamental lack of rationality among Asians and Africans. The assertions are laid on so thickly, and with so little attempt to account for this putative irrationality, that it is difficult to avoid the suspicion that Jones, perhaps alone among modern "miracle" historians, holds some deep prejudices. Africans are dismissed, in a brief discussion, as having had no significance in history. They are characterized in a way that suggests, not very subtly, that they are closer to the beasts than are other humans.

> In Africa man adapted himself to nature. The hunter felt part of the ecosystem, not outside it looking in with wonder, and definitely not above it and superior. After all, there were large carnivores who sought man as a prey. The most evocative symbol of this ecological oneness may be the honey guides . . . birds commensal with man. They fly, chattering loudly, ahead of bands of hunters, leading them . . . to the tree hives of wild bees and feeding on the wax after the men have broken open the hives. (154)[117]

This is a Tarzan image. Most Africans were farmers, not hunters (we discussed this previously). The scavenging house sparrow is also commensal with humans in the suburbs of London, but Jones speaks of "commensalism" only for Africans. (Ethology, as we will see, is evoked somewhat similarly for Asians.[118])

> There was scant incentive or pressure for development or invention. . . . Any pressure . . . was apparently offset by the use of slaves instead of the improvement of methods. . . . Otherwise wealth was spent on luxury items. . . . [There was] pervasive insecurity as a result of conflict and slave-raiding. (155–156)

Africans thus were noninventive, were aggressive, were lovers of luxury, were slavers and slaves. And also hunters close to the beasts they hunted. So: it "is not clear what indigenous developments were possible" (156).

Asians do not think logically. There is "relative absence of the empirical enquiry and criticism of the Graeco-Judeo-Christian tradition" (161), and "lack of a crisp tradition of logical debate," which may explain the "failure" of Asian science" (162). "The notion of a consensus in interpreting nature may have seemed absurd" (162)—that is, Asians may not even have had an ability to conceptualize scientific verification, to distinguish empirical truth from falsity. They were uncreative: "Asian institutions suppressed creativity or diverted it into producing voluptuous luxuries" (231).

Asians have attitudes and values that inhibit progress. "Oriental philosophies [emphasize] emotions, values and cosmologies," at the expense of empirical thought (161). Orientals are lazy (163). They (like Africans) have a "love of luxury" (170), and they love to buy things like "kingfisher feathers ... precious stones ... drugs no modern pharmacopeia would own" (164). They have a "servile spirit," their armies lack "tough" petty officers (167). They are, as a rule, introverted, inward looking; they are "increasingly immobile societies undergoing 'curious experiences,' " given to self-imposed "isolation" (170), lacking an urge to explore (168, 177, 203, 231). They are given to senseless warfare (169, 188, 197), do not have a written legal system (164, 188, 197), and do not have a concept of political boundary (167, 194). There is much thievery and piracy (189, 199, 209, 229–30).

In most judgments of this type (of which I have given only a sample), Jones is referring to Asians in general and through all times in their history.[119] Some judgments are more specific. Islamic society was for a time somewhat innovative, borrowing technology from other societies, and preserving ancient Greek sciences. (This is a standard Eurocentric belief: Arabs preserved Greek science during the Dark Ages and then handed it back to Europe for further development, something like an intellectual left-luggage office.) The Ottoman Empire (the single real example of Islamic culture discussed by Jones, with the imputation that it stands as symbol for Islam as a whole and at all times) stamped out original thought. It produced unreason, backwardness, and retrogression, a "mist of obscurantist thought" (183). Ottomans didn't even know "the elementary facts of geography" (184) and couldn't even make decent maps (179). (Jones himself doesn't know much about geography and maps.) Rulers were often "degenerates," "drunkards," "mental defectives," "lechers," ruling with despotism and terror (186–187), their "philosophy" being theft and despoilment, against which there was "no legal shield" (187–189).

Indian society was socially and psychologically "frozen" (192), with values that were deleterious to economic progress. Religion was invoked to sanction all acts, but the advice of religious counsellors was "malicious or random" (195). The Mughal rulers were, like the Ottomans, degenerates, running society for their own benefit, given to "voluptuous selfishness," harems, jewels, menageries, intrigues, treason (196). The state was purely predatory. Technology was "almost stagnant," not even copying from abroad (199). "No written legal code existed," says Jones in a quite bizarre misstatement (197).

China was technologically inventive and innovative until the early Middle Ages, when progress stopped. There was thereafter a "retreat" (203); some skills were even forgotten. Chinese became "inward-looking"

(203, 216, 220). China "backed away" from technology, from trade, from exploration. Technological development stopped even in agriculture, and only the cultivation of new land, with irrational cutting of irreplaceable forests, and the fortunate, timely, arrival of New World crops such as maize and sweet potato, saved the Chinese temporarily from disaster, although cutting down these forests was "one of mankind's greatest acts of ecological stupidity," stupidity which led to "soil erosion, gullying, silting and floods" (213). (Apparently the deforestation of Europe and North American was not "stupid.") Peasants were given to "envy and suspicion" (206), were stupid as farmers (212–217), and were stupid also in preferring "maximal reproduction" over "affluence" (218). The state was despotic, a "revenue-pump" for the rulers, providing no services (206). There was (again) a love of luxury, an attitude of "empty cultural superiority" (205), a corrupt, venal, parasitic ruling class. Chinese had "anti-social customs" (7) and were diseased.

Ethology, the science of comparative animal behavior, is invoked repeatedly by Jones in his characterization of Asians and Africans. Oriental courts were given to displays of "the dominance relationship" (109, 209). Multiple wives had "ethological significance," as also, perhaps, did the "amassing" of slaves for "display purposes." "Great attention was paid to submission symbols, kneeling, prostration, the kotow" (209). In India, a "similar calculus . . . underlay human demographic behavior and veneration of the cow" (19). Recall the "commensalism" in Africa.

These beliefs about the intellectual and moral inferiority of non-Europeans are old colonial-era prejudices, which Jones merely recites. The fact that *The European Miracle* is taken seriously by some historians is partly, I think, a reflection of the degree to which these old prejudices still lurk within our scholarship as implicit beliefs.

Many other historians of the "miracle" try to insert European rationality at or near the base of their explanatory theories. Michael Mann can be taken as a typical representative of the genre. We have already discussed his contrived little theory about the marvelous Iron Age Indo-Europeans of the northern forests, uniquely individualistic, aggressive, freedom loving (all notions from nineteenth-century and earlier scholarship, none grounded in solid evidence for Europe or, comparatively, for non-Europe). Mann moves smoothly to ancient Greece, which he sees as the product of this earlier Indo-European root rationalism. The Greeks invented most things rational, from science to democracy to ethics (a respect for humanity). Other Mediterranean peoples contributed rather little by comparison with the rational Indo-Europeans who swept down from the north. (Even the Roman influence, which subjugated the

European peasant, is seen as vaguely "un-European."[120]) History then moves northwestward as northern Europeans clear the forest and move what Mann calls the "leading edge" of European society forward toward what seems to be its foreordained goal, Britain.[121] Early in the Middle Ages, around A.D. 800, a newer stage of rationality emerges, combining the old culture and the new Christianity. Europeans now demonstrate their "rational restlessness" and their unique inventiveness, producing a revolutionary agricultural system and much more. The explanation is the ancient rationality, the newer Christian rationality, and the marvelous European environment.[122] This, for Mann, is the kernel of the European miracle. The only part of this construct that requires further comment is the theory of a medieval technological revolution; this we discuss in the following section.

Lynn White, Jr., contributed a very influential form of the rationality doctrine with his theory about the unique technological inventiveness of medieval Europeans, a theory we discuss in the next section. John Hall contributes another version, much like Mann's except for its emphasis on the invention of rational politics by Europeans and its tendency to follow Eric L. Jones in various negative judgments about Asian history. Space does not permit me to review the way some other present-day historians invoke, and use, the theory of Western rationality as grounding for their theories about the European miracle. I should simply note that some of them are Marxists and near-Marxists. Perry Anderson, for instance, argues from ancient Greek rationality to European medieval progress. So does M. I. Finley. Robert Brenner contributes a very different Marxist rationality theory: there was *no* rationality until, quite suddenly, capitalism appeared among the English yeoman farmers, who promptly became amazingly inventive and started a technological revolution that has not yet ended.[123]

Technology

Among all of the narrow-minded ways of looking at history, technological determinism is the one most congenial to Eurocentric tunnel vision. It has the appearance, the illusion, of cold-blooded scientific fact. "X was invented *here*, on *this* date, and produced *these* effects." In talking about matters of technology, one can deny that ethnocentrism enters the picture: "this is hard fact, indisputable." And the significance of technology seems equally indisputable: a new tool does produce more food, a new weapon more casualties, whereas a new social form *might* have historical significance; then again, it might not. But technological determinism gets its greatest strength from the error known as "telescoping history." When we travel, mentally, back to medieval Europe, we pass backward through the eras in which Europeans clearly were technologically superior to everyone else, and so we tend to expect that superiority to have been the state of things at all prior times. But Europe advanced technologically beyond Asia and Africa mainly after the beginning of the industrial revolution. Europe did not even begin to forge ahead of other civilizations in technology or science until the seventeenth century or even later. By telescoping history we imbue the Middle Ages with the marvelous technological attributes of modern Europe. It is then but a small step to the conclusion that, since technology is so *obviously* a powerful cause in history, and since Europe has *always* been so technological, *here* is the root of the European miracle.

But tools do not invent themselves or reproduce themselves. If you invoke technological determinism, you must not only show that the technology appeared and had such-and-such effects, you must also explain why it was invented, and by whom. In nearly all (I think) technologically deterministic arguments made as part of some explanation for Europe's historical progress, technological arguments end up being arguments about the inventors, not the inventions. They end up in some Weberian claim that Europeans are more inventive, innovative, "rational," than non-Europeans. Technological determinism then differs from other kinds of tunnel-historical theories mainly in its claim that rational Europeans moved their society forward by inventing new technology, rather than doing so by inventing new political systems, new forms of social organization, new religions, or whatever.

Claims are made by Eurocentric historians that technology was a prime mover in all epochs from the post-Neolithic on down. Nowadays it would be hard to ignore the evidence that in ancient times the greatest technological innovations, in agriculture, transportation, and other spheres, entered Europe from the south and east, so technological arguments tend to begin with the Middle Ages. There are, however, exceptions. Michael Mann, as we noted previously, has revived the old notion that the invention of iron somehow revolutionized European peasant society in a way it did not do elsewhere, that it produced a wondrously inventive, acquisitive (etc.) type of society, rooted in iron-plowing Indo-European–speaking peasants, and he claims, in parallel, that Oriental despotisms of the ancient Middle East held back technological inventiveness, where peasants were supposedly held in tight control by the political system that irrigation technology necessitated. But this theory is contradicted by a great deal of evidence. So much is known about technological progress in the Middle East during this period (from the late Neolithic to the "Iron Age" and beyond) that the notion of noninventiveness is unreasonable. But iron working was not invented in Europe. Probably it came from the Middle East, but there are hints from farther afield (including West Africa). Plows were used in Middle Eastern agriculture, irrigated and nonirrigated, as well as in European agriculture. Plows may have been used earlier in China than in Europe. So the technological argument put forward by Mann (and others) is unsound and the psychological and social deductions are fallacious.

Many other "miracle" historians are prepared to start with the ancient Greeks and define these folk as the original uniquely inventive Europeans, but the Greeks did not contribute more to technological progress than their neighbors did to the south and east, so arguments of this sort must revert to the abstract idea of "rationality," thus, of potentiality, not actuality, a point we have already discussed. The Romans, in turn, were moderately inventive, but so were *their* neighbors. In the European Dark Ages, as everyone agrees, technological progress was hardly impressive, and was in fact going on at a healthier pace in parts of Africa and Asia. I suspect that the majority of historians today are not willing to go along with Michael Mann and others who claim to find a unique technological progress in Europe prior to the Middle Ages, although this view was quite widespread until a few decades ago. Most would disagree with Eric L. Jones, in *The European Miracle*, that Europe has always been "a mutant civilization in its uninterrupted amassing of knowledge about technology" (p. 45), since there is nothing unusual about this process in human society.

It is with the Middle Ages that technological arguments come to prominence. It is here that the notion of a technological "takeoff to modernity" is most frequently encountered among historians of the European "miracle." The common argument form is about as follows: Northern Europeans indeed were rather backward folk until late in the Dark Ages, the early Middle Ages, when a kind of awakening occurred, and northern Europeans now suddenly emerged as the people we know—the most dynamic, most progressive, most innovative, most rapidly rising (etc.) people in the world—and the key to this rise was a set of early medieval technological innovations, based mainly, these historians claim, on *European* inventions, neglecting non-Europe and thus presenting us with a typical sort of Eurocentric tunnel history.

Lynn White, Jr., is an American historian whose book *Medieval Technology and Social Change* presents this kind of technologically deterministic tunnel-historical argument in perhaps its purest form. This book, published in 1962, has been highly influential; its arguments are built in to the theories of many "miracle" historians, including Mann, Jones, and Hall (who not only cite the book repeatedly in their writings but rely on it for many technological arguments they invoke in support of their theories). The book is an effort to show that technological invention and innovation was the central cause of the rise of Europe during the Middle Ages. White lists a series of supposedly European inventions, and shows, for each in turn, what marvelous effects it had on European history. Here, for starters, is an example. The early medieval invention of the iron stirrup had a "catalytic . . . influence on history." It permitted a new form of mounted warfare, created the phenomenon of the medieval knight, and produced feudalism (knights became manorial lords). And so, "The Man on Horseback, as we have known him during the past millennium, was made possible by the stirrup."[124]

But White's crucial arguments concern productive technology, and particularly agricultural technology. He claims that an agricultural revolution occurred in Europe (or rather in northern Europe) in the early Middle Ages, and this quite revolutionized European society and became a large part of the explanation for modernization and the rise of capitalism. He believes that three European inventions were the crux of the matter: the heavy plow, the horse collar (and therefore the use of horsepower), and the three-field rotation.

The heavy plow, pulled by teams of (typically) eight oxen, is assigned, by White, a tentative central European origin in the sixth century; it then diffused quickly throughout northwestern Europe, and "does much to account for the bursting vitality of the Carolingian realm in the eighth century."[125] White is correct in calling attention, as others before him have done, to the importance of the heavy plow as an agricultural innovation in the wetter and colder parts of Europe. It was highly advantageous in opening the deep, heavy soils of the North European Plain and the deep-working of soil was critical in view of northern Europe's moist weather. It was in fact necessary for the areal spread of farming into some of the wetter soils. Therefore, the heavy plow had much to do with the overall increase in medieval agricultural production. But the heavy plow was not in fact invented in Europe in the Middle Ages. Plows pulled by teams of 24 oxen were used in northern India before the time of Christ.[126] Southern Europe used lighter plows, because the soils were generally lighter and drier, but the technology was not significantly different. It seems certain that the heavy plow was either diffused into northern Europe from elsewhere or was a local adaptation of a widely used tool-form. So this is not a European technological revolution. And the effects of the innovation in northern Europe probably should be attributed to the social forces that led to the introduction of the heavy plow, not to the technology itself. We know, for instance, that the growth of feudalism led to massive expansion of cultivation, partly at the incentive of the lords, who wished to increase their wealth and power by expanding their demesnes, partly as a reaction by peasants to the persistently increased demands from the lords for surplus product, increments of demand which could not be met without endangering the subsistence of the peasant family unless the peasants found a way to increase their total production. It appears that whatever historical effects are attributed by White to the plow should rather be attributed to feudalism as a social system. And the effects claimed by him are indeed quite miraculous.

Thanks to the adoption of the heavy plow, says White, there was a tremendous growth in population. Then there was a changeover to the "open field" system of cultivation and this led, according to White, to the invention of "communal patterns" of human cooperation (as though communal life had not been known previously and was unknown elsewhere). This was a social revolution, a "reshaping of peasant society." It was "the essence of the manorial economy," ignoring the fact that the manors were owned by lords, not villagers.[127] More important still, there came then "a change [in] the northern peasants' attitude towards nature *and thus our own*" (my emphasis). Why? Because many families had to collaborate to mount one plow team, and so their allotments now were proportional to their social contribution, not to their needs. ("No more fundamental change in the idea of man's relation to the soil can be imagined: once man had been part of nature; now he became her exploiter."[128])

But all of this is nonsense. Neither the technical arguments nor the social deductions make any sense. The Domesday Book gives household-to-plow-team ratios of between 2:3 and 3:5.[129] The open-field system seems to be quite old and was widespread in Europe and known in Asia and North Africa in early times.[130] Northern villages, using big teams and heavy plows, were no more cooperative than southern villages, using light plows. Communal ownership of fields in some societies implied greater cooperation than that found in the European open field system. The manorial economy was a social system, not a technological invention. And so on.

White's second revolutionary advance is labelled "the discovery of horse-power."[131] Horses had been around for some time, of course. The essential innovation, for White, was the modern horse collar, which he thinks was probably developed in the Occident some time before the ninth century. According to White, the horse collar transformed agriculture and grain transport in northern Europe by permitting horses to replace oxen in pulling plows and wagons. Horses pull about the same weight as oxen but do so about 50% faster. From this fact White draws awesome conclusions. There was, he says, a large increase in agricultural production. Commerce became intensified because transport by horse-power was vastly cheaper than by ox power. Villages became much larger, almost townlike, because now there could be a larger radius of travel from home to field. The enlargement of villages yielded the "virtue of a more 'urban' life," permitting the village to have a church, tavern, school (now boys "could learn their letters").[132] And now there could be "news from distant parts." A transformation, overall, of profound importance. It "urbanized" the village, giving peasants the "psychological preparation" for "the change in Occidental culture from country to city."[133] All this and more from one innovation: the horse collar.

But the horse collar was widespread in Eurasia from an early date, and probably was invented for harnessing not horses but camels.[134] And the presumption that horses held advantages over oxen in plowing and transport is widely disputed: the horse was more efficient, but more costly in upkeep, and generally required that cropland be devoted to feed crops. In England the horse did not replace the ox. Village size had nothing to do with horsepower. In many parts of the world where horses were not used, villages were much larger than they were in northern Europe. In countries like China, long-distance grain transport was often by canal, much more efficient than horse-drawn wagons. And so, again, neither the technical arguments nor the social deductions make any sense.

Finally, White attributes equally marvelous effects to the introduc-tion of the three-field system. Part of this argument is familiar to every European schoolchild, who learned that the three-field rotation was a great advance over the older two-field system because (mainly) it reduced the proportion of the land in fallow from roughly 1/2 to 1/3. But White adds a cornucopia of additional blessings. Oats could now be planted widely, hence there was greater use of horsepower. In a section of *Medieval Technology and Social Change*, "The Three-Field Rotation and Improved Nutrition," he claims that the three-field system somehow permitted farmers now to grow legumes, and this vastly improved the European diet, which, in turn, "goes far towards explaining . . . the startling expansion of population, the growth and multiplication of cities, the rise of industrial production, the outreach of commerce, and the new exuberance of spirits which enlivened that age." In short, says Lynn White, Jr., "the Middle Ages were full of beans."[135]

But none of this (least of all the pun) can be taken seriously. There is no basis for White's argument that population was held down by an unbalanced diet (overloaded, he says, with carbohydrates, undersupplied with proteins). Farmers using the two-field system were not protein-starved, because legume cultivation long antedated the three-field system, grains also contained proteins, and fruits, animal products, and so on, were widely consumed. The three-field system was *not* a technological revolution. First of all, even more intensive rotations, including fallowless systems, were in use in many parts of the world long before the Middle Ages, and I suspect were common in parts of Europe (such as portions of the Po plain) where deep, nutrient-rich soils and good soil–water relations were found. Second, the two-field system was preferred, and was not supplanted, in many ecological situations, and in areas where fallow was needed for grazing. The picture is complex, but the generalization is clear: the three-field system was neither a technical revolution nor a fountainhead for social change. And it had close relatives in other continents, so it cannot be said to have been something uniquely European.

In *Medieval Technology and Social Change*, White says little about the *causes* of the revolutionary technology. He does so elsewhere, however, and we find that his basic argument is really quite Weberian.[136] Technological determinism dissolves into an ideological determinism, focused on the inventiveness—the rationality—of Europeans. This White attributes, basically, to religion: in part to "the Judeo-Christian teleology," and in larger part to "Western Christianity." The former underlies the European's unique "faith in perpetual progress," which leads to a faith in the virtues of inventiveness and technology.[137] The latter produces "an Occidental, voluntarist realization of the Christian dogma of man's transcendence of, and rightful mastery over, nature"—that is, the separation of Man and Nature. To a Western Christian, nature is inert, valueless. It is blasphemy to "assume spirit in nature."[138] This gives the Western Christian a desire and right to manipulate, use, transform nature—to invent new technology. The errors here are glaring. First, there is ignorance of the teachings of religions other than Christianity. Second, White resurrects the ethnocentric myth that ancient pagans and followers of modern non-Christian religions somehow are unable to fully separate Man from Nature and share the primitive view that spirits reside in all things. Actually, medieval Christians *did* perceive spirit in all things. (It was God's way.) They did *not* treat humans as absolutely separate from nature. This dualism is a product of modern times, mainly of the commoditization of nature with the rise of capitalism. Dualism is a doctrine of post-Cartesian thought, not something residing deep in "Western Christianity." Medieval Christians believed that the world is one, a plenum, an uninterrupted chain of being. Medieval folk did not, in general, believe, as White claims they did, in "perpetual progress": they believed in the Fall, and they believed, on the whole, that Creation had been perfect and complete. White is simply telescoping history in order to

claim that modern European attitudes about technology and technological change are actually basic and ancient attributes of the culture of
Europeans, and only Europeans. His central thesis is basic tunnel history:
Europeans are uniquely inventive, so they create unique technology, so
they progress.

I believe that all of the claims for European technological superiority
in the Middle Ages can be countered, and eventually disproven, in much
the same way. I think it will become clear that all of the civilizations of
the hemisphere were inventing new technology and sharing it around;
the "sharing" being the usual mechanisms of diffusion as they operated in
two spheres of reality: agricultural techniques passed from farmer to
farmer and other techniques (including some of those involved in
commercial agriculture) moving through the mainly urban network of
trade and transportation, or occasionally (as with some military
technology) moving around the map along with conquest. Thus I see a
rather even building up of technology in Europe, Africa, and Asia. It was
only after 1492, with its utterly revolutionary consequences, that
European technology acquired the beginning of an edge over Asian and
African. Did medieval technological evolution produce, as cause, the
major medieval social changes? I leave the answer to that question to
others. My purpose in this book is not to defend any one theory of social
change, merely to show that the processes were not uniquely European.

In Chapter 3 I will discuss this matter further, showing how
technological and economic surfaces emerged in medieval Europe, Africa,
and Asia, and that processes of change in all three continents were—as to
technology and economy—quite similar.

Before we leave the subject of technology, something must be said
about one fascinating issue: what I will call "the China formula." Since the
end of World War II a great deal has become known about the history of
technology in China, much of it through the work of Joseph Needham and
his associates.[139] In prior times, the European miracle theorists were prone
to ignore Chinese technological achievements. (Weber did so less than
most.[140]) The usual pattern was to admit Chinese priority for some
innovations but claim that, overall, the main technological advances took
place in Europe and the Mediterranean. Where there were truly notable
Chinese inventions, it was often claimed that the Chinese invented these
things in a ludic vein whereas the Europeans put these things to work.
(The paradigm, of course, is gunpowder, which the Chinese were supposed
to have invented merely for fireworks.) One could bundle such propositions into a classical diffusionist model of the Chinese: they had, in prior
ages, progressed somewhat, but then they slowed to a stop, starting up
again only when Europeans brought new ideas.

It is now known beyond doubt that Chinese technology was on a par with that of the western part of the Old World, in some ways superior to it and in other ways inferior, during and before the Middle Ages.[141] This new knowledge is devastating to many of the European miracle theories, those that claim that ancient and medieval European progress in technology was a crucial cause of the "miracle." (If the Chinese were doing the same things at the same times, the argument about Europe's uniqueness just crumbles.) The result has been a general modification of many of the miracle theories to take these newly known facts into account. Typically, the formula runs as follows.

1. "If indeed the development of technology in medieval China forces us to ask why China did not, then, have its own miracle, we can at least assert that China was the *only* civilization outside of Europe about which such questions arise." In other words, European superiority over everybody else is not put into doubt. This is convenient for those, for instance, who want to show that India, Africa, the Middle East, and so on, had no potential for development.

2. "Whatever technological advances took place in medieval China, the important point is that they *stopped*." In the formula, this argument is expressed in different ways for different spheres of technology, but the central argument is fairly standard: *something* characteristic of Chinese medieval culture forced it to cease developing and so to *stagnate*. In other words, what is plugged in here is the old doctrine of Oriental stagnation. Most typically, Weber is used to make this point: just about all of the Weberian claims about the reasons for Chinese lack of progress are regularly paraded at this point. Some historians balk at using Weber's thin argument about the stultifying effect of Confucianism. Others prefer not to notice Weber's ethnocentrism when he describes Chinese personality traits. But Weber's arguments about the city, landowning, bureaucracy, and empire are still quite regularly employed. The Chinese city was not "free" and did not have a real bourgeoisie. Chinese landowning was not close to private property. The Chinese bureaucracy and the Chinese imperial state were not "rational" and so held back the society from progress.

One can respond to this argument in two steps. First, as Purcell has pointed out, the important question really is how and why these Chinese advances *happened*, not how and why they *stopped* happening (if indeed they did stop). In other words, historians must explain how China came to be such a technologically innovative society that it outstripped other civilizations in many spheres of technology for many centuries. Weber doesn't help in this process one bit. The whole Weberian scheme is an explanation for stagnation, and what we are talking about is impressive progress, not stagnation.

The second step in the argument requires a focus on the precise period when, according to European miracle historians, the Chinese advance is supposed to have stopped. According to Elvin, broad-spectrum technical advances in China ended after the early fourteenth century. China at this time had perhaps the most advanced and most highly commercialized agriculture in the world. Chinese industrial technology was unexcelled in such fields as textile manufacture. Mechanical clocks were well known.[142] Chinese merchant ships were plying throughout Southeast Asia and into the Indian ocean. Chinese guns were unexcelled. Canal technology was impressive. And so on. Broadly, the changes associated with the rise and travails of the Ming dynasty are associated with the slowing of technical advance, although advances continued to take place in some spheres of technology (shipbuilding, cannons, printing with movable metal type— invented circa 1400, probably in Korea[143]—and much more). But Europe at this time was not experiencing a major technical advance either. After 1350 Europe stagnated both economically and technically. There is little evidence that European technology was advancing prior to 1492. The Renaissance was not a technological revolution, as historians have long realized.[144] After 1492 important European advances began again in some technical spheres (notably shipbuilding). Whether truly revolutionary technological change really began before the eighteenth century is a matter of contention among European historians.

What does this say *comparatively* about China? It suggests that there is no problem of "stagnation." There was, instead, a slowing of progress during two centuries, in a scenario well known in all human cultures, because uneven progress is the norm. It suggests, rather, a problem that Third World historians focus some attention on but European historians tend to ignore. As a result of the European commercial expansion in the sixteenth century—we argue in this book that the year 1492 was the real birthdate of this process—aggressive European merchant communities began competing and trading with the Chinese in various places, notably Manila and some South China ports. Why did the Chinese merchant community not assert its dominance in this trade? Why did the commercial advantage steadily move in the direction of Europeans, not Chinese (and other Asians)? In other words, there is no stagnation to be explained. There is, rather, a problem as to how and why Europeans gained substantial control of long-distance trade in Asia after about 1600. Anticipating the discussion of later chapters, I will comment simply that this process does not reflect any internal cultural "blockages" in any Asian civilizations.

Rather, it reflects the tremendously rapid increase in the profitability, scale, and organization of European enterprise overseas after 1492, which nobody else could really take advantage of, because of the constantly increasing flow of New World bullion into European mercantile coffers (a point we will discuss in some detail in Chapter 4).

During the first half of the fifteenth century, a number of Chinese fleets of combined military, diplomatic, and merchant character were sent into Southeast Asia and the Indian Ocean. Some of the ships held hundreds of men; some were so large that vegetable gardens were placed on the decks. We will discuss the significance of these great voyages in the next chapter; my point now is to notice the way European miracle historians try to accommodate these facts into their theories about Chinese nonprogress. The standard comment is: "But they stopped." In fact, the last voyage did end around 1440. But the purposes of the voyages had been accomplished. Chinese merchant ships continued to trade in Southeast Asia. Nothing stopped. It is true that the emperors banned Chinese shipping for a time (mainly the early sixteenth century), but such bans were a means of extracting bribes (in effect easements) and they were not, in any case, enforced. For the "miracle" historians, the fact of the imperial ban is proof that China had stopped progressing, and that only Europe contained the potential for the "miracle." This does not accord with the facts.

Chinese may have invented guns; in any case, they were as advanced in firearms technology as any other culture down to the end of the Middle Ages. Typical of the way European miracle historians deal with this sphere of technology (as with others) is Carlo Cipolla's account of the process in his well-known and influential book *Guns, Sails, and Empires: Technological Innovation and the Early Phase of European Expansion, 1400–1700*. He concedes, to begin with, that "Chinese guns were at least as good as Western guns, if not better, up to the beginning of the fifteenth century."[145] But when the Portuguese appeared off Canton and fired off their cannon in salute a century later (1517), the Chinese had, magically, retrogressed:

> The roar of European ordnance awoke Chinese ... to the frightening reality of a strange, alien people that unexpectedly had appeared along their coasts under the protection and with the menace of superior, formidable weapons. ... How to deal with the "foreign devils"? To fight them or to ignore them? To copy and adopt their techniques and give up local habits and traditions or to sever all contacts with them and seek refuge in a dream of isolation? To be or not to be? ... a dilemma that was tragically unanswerable.[146]

Thus does rhetoric replace fact in the mythology of Asian stagnation and the European miracle.

The "Chinese formula" seems to be losing credence. But I have no doubt that an "Indian formula," an "Islamic formula," an "African formula," and so on, will be contrived to shore up Eurocentric history as we learn more about Indian, Middle-Eastern, and African technological development prior to 1492.

Society

In Chapter 3 we will take a look at medieval European society and compare it to medieval Asian and African society in terms of those social categories that seem to be important in the process of change toward capitalism and modernity. Among these categories (forms, facts) are class, the state, landholding, trade, and urbanization. Our present task is the more modest one of criticizing theories that claim that one or another social category somehow led Europe to rise above all other civilizations before the end of the Middle Ages, before 1492. For this task we can select a few of the categories that present-day historians of the European miracle claim as favored candidates for this role of historical leadership, of causal motor propelling the so-called "take-off to modernity." The favored ones seem to be the state, the church, class, and the family. Most frequently these categories are, themselves, explained in terms of supposedly more basic forces: European rationality, European technology, European demographic uniqueness, or European environmental superiority. These have been discussed at sufficient length already. The following discussion, therefore, will be brief.

State

A typical sort of argument begins with a model of the modern European nation–state, the kind of medium-sized, well-integrated, moderately democratic state that emerged after the rise of capitalism, after modernization, after the French and American revolutions. Nobody denies that this form is unique. But European miracle historians tend to make one of two claims about this state form: either it appeared very early in European history, early enough to play a causal role in modernization, or it was somehow immanent in European culture, a state form that was created, naturally and rationally, by freedom-loving, individualistic, antidespotic Europeans of medieval or earlier times. Sometimes both claims are made: a reduction to European moral rationality and, at the same time, a holistic argument at the level of the state itself.

Eric L. Jones, for instance, does precisely that in *The European Miracle*. Europe's medieval political system, says Jones, was near "the heart of the European miracle."[147] Here the argument combines the traditional prejudice about Europe's innate and unique love of freedom with some very odd environmentalistic arguments. First the latter. Jones takes the familiar argument that Europe in early times had a number of fertile core areas, each of which became the hearth of a regional culture, and he turns it into a curious argument to the effect that Europe's ecological core areas naturally led to a pattern of medium-sized states, unlike, he says, the "imperial" pattern of Asian societies. Empires, in Jones's view, are innately despotic, and innately given to interfering with the development of the economy. Europe's geography saved it from this fate. If a number of Europe's modern states have ecological "core areas," most do not, and rather few of these core areas became states. The model is environmentalistic and invalid. In any case, comparable core areas are found in many other parts of the world. Medium-sized, separate units of productive agricultural environments are, for instance, typical for Southeast Asia, including the rather discrete mainland cores, like the Irrawaddy and the Chao Phraya basins, the middle Mekong, and the Red River, and even more so the insular region, with cores in Sumatra (two or three), Java (three), Bali, Lombok, Sulawesi, Luzon, and so on. India and China too are moderately dissected in this way. Distinctive cores in Africa include the Middle Niger, Chad, and Congo, among others; in western Asia, Mesopotamia, the Iranian Plateau, and so on.[148]

Next, Jones builds the image of these natural European *Ur*-states as somehow naturally forming into what he calls "a states system," a grid of roughly equal-sized independent states. These, in the Middle Ages, acted toward one another as though they were Hegelian individuals, competing yet cooperating, and marching forward, together, as a kind of members' club, toward modernity. Thus Jones discerns deep in the Middle Ages what amounts to the modern nation–state and the League of Nations. Telescoping history, he imbues the medieval European polities with all the virtues of the modern state: they provided public services, they encouraged the free development of the economy, they were incipiently democratic, and so on. All of this because of their environmental basis and because freedom-loving Europeans lived in them. In reality, the kind of state, and state system, that he describes does not appear until roughly the seventeenth century. Jones finds political characteristics of already-modernizing Europe, falsely claims that these were present long before—some, being environmental, were always present—and so claims to have proven that Europe's modern democratic political system was, somehow, always present in Europe (and nowhere else). In fact, medieval

Europe was a hodge-podge of semisovereign feudal political units, a map so confused that states as such were not always identifiable, and with few of the modern state characteristics attributed to these polities by Jones.

But to understand the absurdity of this model of political history, we must notice the way Jones talks about the supposedly barbaric, despotic states of non-Europe. In non-Europe, a natural irrationality combines with environmental disadvantages to produce the opposite of the wonderful European state and system of states: the huge Oriental empire. Why did Asia not develop politically as Europe did? The political "infantilism" of Asia is explained by Jones in terms of (1) a psychological deficiency, consisting of irrationality in matters of intellectual vitality and innovativeness and a sort of moral failing in attitudes relating to the desire for progress, resistance to domination, will to forego animal pleasures, and the like, and (2) an inferior natural environment. The effects of these failings (and some lesser ones), and thus the effective reasons for Asia's political nondevelopment and nonprogress, are (1) uncontrolled population growth, and (2) bad government, "Oriental despotism" or, as Jones prefers to label it, *empire*. Throughout *The European Miracle*, the main argument supporting the theory about how Europe's system of states favors development is Jones's countertheory about the evil nature of the Asian imperial state. Yet, in the end, the countertheory turns out to be as empty as the theory. Jones does not, in fact, give any credible explanation as to why such large empires exist in Asia and not in Europe. But neither does he give a credible argument as to why large states are worse for progress than medium-sized or small ones. (And he is blind to the very real progress that occurred in Asia, and to the many nonimperial states in Asia.) When we examine Jones's theory about Asian empires, then, we find that it has nothing to do with size or kind of state and everything to do with a conception about the basic nature of politics in non-European societies. This is nothing more than the old idea of "Oriental despotism." Asians, and indeed all non-Europeans, naturally suffer nasty, despotic, capricious, irresponsible, evil governments. Only Europeans understand and thus enjoy freedom.

This is not an atypical theory among historians who wish to argue that European political processes and forms were causal forces leading to a medieval "miracle." The basic form of the theory is traditional. It consists mainly of three propositions: Europeans have an old, deep-seated rationality that impels them toward modern, implicitly democratic, political forms (or, less directly, toward individualistic economic behavior and thus a preference for a minimalistic state); non-Europeans suffer gladly and naturally a despotic, imperial form of state, sometimes powerful but always arbitrary; and Europe's natural environment gives political modernization one or another sort of unique boost (ecological cores, fertile soil, capes-and-bays accessibility, or whatever). Other propositions figure in most arguments, but these three seem to be the most widely used.

They combine with a methodological principle of great significance: whatever causal forces were at work, they were at work *early*—that is, they set the process of modernization in motion long before the end of the Middle Ages. This, of course, is the crux of the matter. In the present book I argue that modernization (or whatever you want to call it) was indeed going on in medieval Europe, but not *only* in medieval Europe. Thus the events after 1492 did not start modernization but enabled European changes to increase in magnitude and effect and eventually produce a uniquely "rising" society. Therefore we argue from the uniformitarian hypothesis that Europe had no potential greater than that possessed by non-Europe during the Middle Ages. The fact that there were forces pushing toward democratization and the modern state in medieval Europe tells us nothing about European uniqueness.

John A. Hall uses the basic Jones model, emphasizing what he calls the "blockages" that prevented China, India, and the Islamic Middle East from developing modern state forms. In brief: China suffered under empire. India suffered under caste (and so had no politics whatever).[149] Islamic-region politics reflected tribalism. Hall describes very old and not really typical forms for each of these civilizations, assuming them to be frozen and permanent, contrasts these with late postmedieval European forms with the inference that these too have been present at least in embryo for a long time, and then builds an argument that non-Europe always lacked the potential for a "miracle," while Europe always had it.[150]

Michael Mann addresses himself to the abstract entity "power" instead of the more visible "state," but his argument is not fundamentally different. Mann, like a number of other "miracle" historians, tries to interpret the political chaos of early medieval feudalism, with its fragmentation of polities and sovereignties, into an argument for incipient modernity. The smallness of these politicosocial units implies, somehow, that they were more "intensive," more meaningful, more pregnant with development potential, than the supposedly vast imperial polities of non-Europe. Mann's notion of "intensiveness" is basically a notion of value, not of fact.[151] In this context we may recall the Magna Carta myth: devolution of power from king to barons is traditionally described as a movement toward democracy, even though the barons in those times were hardly democratic, and poor people preferred to be able to appeal above their heads to the king. (And later the rise of royal power is described as a further step toward democracy, except that you can't have it both ways.[152]) Jean Baechler uses roughly the same model—although, unlike Hall and Mann but rather like Jones, he downplays the role of Christianity in political evolution—with one additional argument: Baechler believes that the medieval European aristocracy (originally an Indo-European "warrior aristocracy") was itself the original source of democracy, claiming, in a curious reading of history, that aristocrats not only dealt democratically (respectfully) with one another, but they also respected the rights of their peasants.[153]

Yet there was no democracy in the Middle Ages: European states were as despotic as any found anywhere else. Just as we noted with regard to the ancient Greeks, political forms found in Europe had counterparts elsewhere. If city–states were (perhaps) the closest to democracy—some of them were republican, and were in essence merchant-oligarchies—one need only note that city–states of many sorts lined the coasts of the Indian and Pacific oceans. Nor was the republican form unknown in Asia and Africa.

Church

Many theories assign one or another causal role to the church as a social institution in the modernization of Europe and the rise of capitalism.[154] (Many other theories assign such a role to Christianity itself, but I will not comment on these.) Of course, the church did have such a role. The problems lie elsewhere. Most of them concern the question whether the church (or churches) provided Europe, and Europeans, with something that was not provided to non-Europeans by their own religious institutions or by other institutions in their own cultures. For instance, the claim has been made that the Catholic church in the Middle Ages unified Europe in cultural terms, to a degree not found elsewhere. But elsewhere comparable or parallel processes were at work. For instance, the much-maligned empires, such as the Chinese empire, provided unity in many ways, while the Islamic religion provided it in the Muslim world. It has also been argued that the medieval church compensated for the political fragmentation of Europe, allowing pan-European development to take place in spite of the lack of a clear pattern of strong states. True. The question is comparative: did the church thereby give Europe some developmental advantage over other civilizations, in which, quite often, there was political unity throughout most historical epochs? I think the answer must be that the church did not give Europe some special cultural quality, lacking in other civilizations, such that Europe could thereby leap into the forefront of social evolution.

Therefore I see no basis in fact for suggestions that the medieval church led to European historical superiority in a way that other religious institutions, in other civilizations, did not do. John A. Hall, for instance, asserts that "Christianity provided the best shell for the emergence of states."[155] Michael Mann claims that the medieval church "encouraged a drive for moral and social improvement even against worldly authority"—a statement that one might question not merely as a comparison, but also as a factual assertion about the role played by the church in Europe.[156] For H. E. Hallam, "the medieval Latin church was the seed-bed of the early modern idea of capitalism."[157] K. F. Werner maintains that "the 'European miracle' [took place] . . . because of the existence of a Christian world dominated in the west by Catholic doctrines."[158] Such statements fail as comparisons with other religions, and other civilizations; but they also fail because the postulated rise of Europe did not take place in the Middle Ages but somewhat later. That Christianity helped in that eventual rise is not at all in question.

It is not necessary to discuss Max Weber's celebrated argument about the role of religion in the rise of Europe, or his celebrated argument about the special role of the Protestant Reformation in the rise of capitalism. Weber of course held the views that I criticized above. But the causal role of religion is not at issue in this book, the concern of which is a uniformitarian view of European and non-European civilizations and their dynamism.

Class

Three kinds of argument concerning class are relevant to our discussion. The first of these is a two-tiered argument for Europe's ancient and medieval historical superiority that can be summarized as follows: (1) Classless (or "preclass") regions and peoples are simply irrelevant in the matter of explaining Europe's superiority and the world's historical progress, because classless societies are necessarily both unprogressive and primitive. (2) Therefore, the real problems calling for explanation concern the question why some class-stratified societies (the European ones) are progressive while others (principally Asian) are stagnant and backward. Africa below the Sahara is then declared sweepingly to be classless and thus irrelevant to world history. This argument is still widely used, in some analytical works (among them Eric L. Jones's *The European Miracle*), in some world-history textbooks (as we noted in Chapter 1), and in some (probably most) modern atlases of world history. (Many of these atlases have *no* maps of Africa for the whole of history from the Upper Paleolithic to 1492![159]) In place of an extended discussion of this issue I will just offer these comments: Africa was not classless in 1492 (a point to which we return in Chapter 3), classless societies are not stagnant,[160] and the historians who make this sort of argument about classlessness quite regularly contradict themselves by declaring that the classless European peoples of ancient and medieval times—the "barbarian tribes," Germans, Celts, Slavs, etc.—were *very* progressive: after borrowing the idea of classes and a few other things from the Romans, these primitive but supposedly progressive, innovative, aggressive, inquisitive, achievement-oriented Europeans than surged forward to modernity.

It is customary and quite proper to examine the class structure of a society in history and ask how the different classes acted to favor or resist change, and in which directions. In the absence of such an examination, we usually get an analysis with a bias toward ascribing causality solely to kings and other sorts of elites. Eurocentrism becomes a problem only if the assertion is made that a given class did something historically efficacious in Europe but that this class did not exist in civilizations outside of Europe. This kind of argument is very common in the European miracle literature. Although especially favored among conservative historians, it is also found quite often among those Marxists who cling to the old formulas about the "stages of class society." Fairly representative of the latter is Padgug's argument that a slave mode of production, in its pure form—with slavery dominant in commodity production and with implications for economic progress, the development of private property, and class struggle—existed in ancient times only in Greece and Rome, and that lack of this feature in Asia accounts in part for the tendency of Asian societies to "stagnate."[161] (Roughly the same argument is then used by some Marxists to assert that the slave-plantation system of later times was not properly capitalist because the workers were slaves, not wage-earners. We deal with this point in Chapter 4.) This issue is important, but I will set it aside after one comment: I believe that those scholars who assert that the slave class was more important in the classical Mediterranean than Asia make a very elementary error, perhaps easiest to do if one is not a geographer. It may just be a question of geographical scale. The Athenian empire was perhaps one-hundredth the size of the Ch'in empire. In small, highly developed regions of China, slavery was very likely as important as it was in Attica or in the plantation regions of Roman Italy.

Jean Baechler, a French historical sociologist, has argued that one uniquely European class, the medieval aristocracy, was the central causal force in the European miracle.[162] He dismisses Africa out of hand, then asks why it is that modernity arose in Europe and not in Asia.[163] Accepting many of the standard arguments as partial explanations, he argues that the most important reason the miracle occurred was the existence in Europe of a *true aristocracy*. Baechler describes an ancient

Indo-European society characterized by its warrior aristocracy. In India, the aristocracy became corrupted, but not so in Europe.[164] He carefully defines the medieval aristocracy and its special social quality, "feudality," in such a way as to distinguish it from a mere landlord class, or a class of landlords with added seigneurial powers. The aristocracy was a band of comrades, equals joined by bonds of feudal loyalty, a democracy in its own right. It did not, he says, hold the political power in society, and this was a key to its unique character. Elsewhere, either the aristocracy was ground under by the (imperial, despotic) polity, or it became corrupted into a caste form, as in India; in such cases there was to be no modernization. The European aristocracy had a special sort of *private* power, one original source of capitalist power. (Baechler contradicts himself here, since in the feudal era the aristocracy *was* indeed the political power. He simply notes that there was a time of political chaos, and passes on to other matters.[165]) In sum, the aristocracy invents democracy and incipient capitalist property. The European peasants are also incipient capitalists from early on. Baechler casts aside the idea that Europe's peasantry was an unfree, oppressed community in the Middle Ages and gives it quite remarkable qualities: The peasant village was self-governed, a kind of "republic," with characteristics reminiscent of urban life.[166] "A peasant is an entrepreneur in miniature."[167] By the fourteenth century this incipiently capitalist peasant is the real peasant of Europe. The peasant village is a little democracy. The peasants are autonomous decision-makers. Baechler simply ignores serfdom, ignores, likewise, the importance of the fact that the lords owned the land, exploited the peasants, and controlled their lives. Feudalism becomes a kind of democratic society in which the aristocrats play a democratic role and the peasants live as free people. Moreover, these peasants, deep in the Middle Ages, had all of the attributes of the capitalist farmer that we associate with the eighteenth century and thereafter: investment, profit orientation, capital accumulation. (And they were smart enough, Baechler says, to avoid the extended family.) All of this is fantasy. It is a simple telescoping of history, pushing the modern world (and particularly the modern capitalist farmer) back into the Middle Ages.

Baechler contrasts all of this with India. His depiction of India is bizarre. India had no aristocracy throughout history, as a result of the caste system, and this is the "deepest cause" for its failure to develop throughout history.[168] (India *did* have a very formidable aristocracy.) Indian society had no political dimension: "The polity . . . . is not a reality in India." Therefore "the sense of identity in India could never be political"—not to say democratic.[169] (Nonsense.) Caste was invented to take the place of the missing polity. (More nonsense.) India didn't have a peasantry, it merely had "agricultural workers."[170] How can this be true? Because real peasants were entrepreneurial decisionmakers, something supposedly absent in India (but in reality no more so than in Europe). Baechler thus uses a mythical India as a counterpoint to his basic argument: In Europe, aristocrats and the free society created by them (and their allies the peasants) were the principal source of the "miracle." But Europe's aristocracy and peasantry did not have these romanticized qualities, and, more crucially, comparable classes existed in many other regions in the same period, as we will see in Chapter 3.

In Marxist theory, the concept of class is part of a larger and more consequential notion, that of class struggle. For all class-stratified societies, said Marx and Engels, class struggle is the motor of progress. Most present-day Marxists consider cultural evolution to be very much more complex than this, but they continue to emphasize the idea, and process, of class struggle. Again we must notice that there is nothing inherently Eurocentric about this concept unless one argues that genuine class struggle, or some phase or form of it, occurs *only* in Europe. But many Marxists argue precisely that way. According to Maurice Godelier, for example, the West displays "the purest forms of class struggle" and "alone has created the conditions for transcending . . . class organization."[171] Probably the most influential recent formulation of this sort of argument is Robert Brenner's theory about the rise of capitalism.[172] Brenner tries to demonstrate that the rise of capitalism prior to 1492 was a result of class struggle, but class struggle *only* in Europe. Then he uses this argument as evidence against what he calls "Third-Worldism," the belief that non-Europe has been of great importance historically and is so (politically) in the present. This conclusion has strongly influenced not only Marxist but also conservative thought in history, geography, sociology, and economic-development theory. Certainly the Brenner theory deserves our attention in this book. And it is not at all a complicated theory.

According to Brenner, class struggle between serfs and lords, influenced by depopulation, led to the decline of feudalism in northwestern Europe. (Brenner does not mention non-Europe and scarcely mentions southern Europe.) In most parts of northwestern Europe, the peasants won this class struggle and became in essence petty landowners, now satisfied with their bucolic existence and unwilling to innovate. Only in England did the lords maintain their grip on the land; peasants thus remained tenants. The peasantry then became differentiated, producing a class of landless laborers and a rising class of larger tenant farmers, wealthy enough to rent substantial holdings and forced (because they had to pay rent) to commercialize, innovate technologically, and thus become capitalists. (Brenner thinks that serfs, lords, and landowning peasants did not innovate, and that towns, even English towns, had only a minor role in the rise of capitalism.) English yeoman tenant farmers, therefore, were the founders of capitalism. Stated differently: capitalism arose because English peasants lost the class struggle. In reality, however, peasants were not predominantly landowners in the other countries of the region; capitalism grew more rapidly in and near the towns than in the rural countryside; and the technological innovativeness that Brenner attributes to fourteenth and fifteenth century English farmers really occurred much later, too late to fit into his theory. More importantly, commercial farming and indeed urban protocapitalism were developing during this period in southern Europe and (as I will argue) in other continents. Brenner's theory is simply wrong.[173]

Family

There is nothing new about the belief that the European family is in some fundamental sense more rational and more civilized than the family types found elsewhere. When the modernization doctrine became dominant, this belief seems to have faded into the background; there was now a near-consensus among social scientists that differences in family type should be strung along the continuum from "traditional" to "modern" or, as a variation, from "folk" to "urban." Traditional families, it was argued, have tended to reflect strong kinship ties at a scale larger and wider than the nuclear family; they have tended to form large, extended-family households; and they have tended to be associated with high birthrates. In the absence of much evidence concerning the medieval European family and household, it was rather widely assumed that earlier European families had conformed to the traditional model, which was seen as the "preindustrial" family, contrasted with the modern, post–industrial-revolution nuclear family, forming a small household, with looser kin ties beyond the household, and with fewer children. This transformation was thought to be associated with what was called the "demographic transition," the process of change from a "traditional" demographic pattern, with high birthrate and high deathrate, characteristic of preindustrial conditions, to a "modern," postindustrial demographic pattern, with low birthrate and low deathrate.

In the modernization theory of the family, the non-European world of underdeveloped countries would go through essentially the same transformation as it modernized. Extended family households would be replaced by nuclear family households. This would instill modern prodevelopment attitudes: people would have fewer children (thus combating overpopulation), and they would tend, now, to think more individualistically and thus entrepreneurially.

All of this was a bundle of assumptions. There was no evidence of a causal link between birthrate and size of household (or strength of kin ties beyond the nuclear family); there was merely a correlation within rich, modern societies: *both* nuclear families and lower birthrates were characteristic of these societies. (One leading demographer pointed out that nuclear families might, in principle, be likely to have more children per couple than extended families.[174]) In addition, the idea that an adult breadwinner would be more entrepreneurial, more prone to accumulate capital, more competitive, and so on, if he (assumed to be male) was working only for his wife and children and not for a large family including his parents, cousins, and other relations, was a tenuous assumption. Large, tightly knit families had proven to be powerful accumulators, for instance, among many immigrant groups engaged in commerce. Why, indeed, should an adult want to work harder only for spouse and children, and not for parents, sisters, and so on? (One can in fact construct an argument to the effect that an extended family with multiple working adults has definite economic advantages in a developing economy.[175]) As the modernization doctrine itself developed, after about 1960, these ideas began to dissipate. It became clear that various forms of social organization can be equally modern, and that the root causes of change and nonchange do not lie in the family structure.

Rather dramatically, in the mid-1960s, the notion of a peculiarly European family pattern was reinserted into historians' common discourse, partly as a result of an influential paper by John Hajnal.[176] As we noted at an earlier point in this chapter, in the discussion of demographic arguments for the "miracle," historical demography was at that time uncovering evidence that preindustrial Europe (or part of Europe) had a lower birthrate than one would expect of a "traditional society." It seemed that Europeans had been marrying later in life than would be expected in a traditional society with Malthusian controls and therefore maximum birthrate. Instead of scrapping the Malthusian model itself, and arguing instead that all human societies actively control their population dynamics by using such tactics as adjusting the age of marriage, these historians began to assert, simply, the uniqueness of Europe. Other preindustrial societies exhibited the "traditional" pattern, with high and uncontrolled birthrates (and therefore superrapid population growth when mortality declined in the "transition"). Preindustrial Europe, in complete contrast, had a rational family system, with rational population control. It was then argued, without evidence, that preindustrial Europe had a higher level of living and a lower mortality rate than non-Europe, from which it was deduced that Europe kept its birthrate down in proper, rational, adjustment to its lower deathrate, mainly using the mechanism of delayed marriage. Those who put forward this argument simply failed to consider the possibility that families in other societies have acted pretty much the same way. Evidence to this effect was beginning to appear from a number of non-European societies. Some societies had much lower birthrates than would be expected from the classic model. Other societies had demonstrated major shifts in birthrate, either upward or downward, in response to changes in economic and other conditions.[177]

Hajnal began his 1965 essay with this flat statement:

> The marriage pattern of most of Europe as it existed for at least two centuries up to 1940 was, as far as we can tell, unique or almost unique in the world. There is no known example of a population of non-European civilization which has had a similar pattern. The distinctive marks of the "European pattern" are (1) a high age of marriage and (2) a high proportion of people who never marry at all.[178]

Hajnal's argument was in some ways very careful and in others very casual. He noted carefully that the evidence for late marriage and (somewhat) low marriage frequency in Europe for periods before the seventeenth century was inconclusive, and that the "fragmentary" evidence for the European Middle Ages suggested a "non-European pattern."[179] He introduced absolutely no historical data for non-European regions, casually comparing historical Europe with twentieth-century non-Europe. The theoretical constraint is very clear: non-European patterns are "traditional" and permanent, hence a comparison of seventeenth-century Europe with mid-twentieth-century Asia or Africa is perfectly acceptable.[180]

This proposition about the unique European pattern of later marriages and lower marriage frequencies was widely and quickly incorporated into the broader theory of Europe's historical "miracle." Without much supporting evidence, the proposition was firmly pushed backward into the Middle Ages. According to Lawrence Stone, writing in 1977, "it has now been established beyond any doubt" that over most of northwest Europe the middle and lower classes "married remarkably late, certainly from the fifteenth century onward. . . . This custom of delayed marriage is an extraordinary and unique feature of north-west European civilization."[181] Here one notices particularly the word "remarkably." This is "remarkable" *only* if there is something to compare it with; yet non-European historical data are not given or even sought. Michael Mann derives the pattern from the Iron Age Indo-European peasant society. Patricia Crone speculates that it is perhaps an ancient Germanic trait. Eric L. Jones thinks that it goes back three or four thousand years. Alan Macfarlane thinks that it has its roots in "a particular amalgam of Christianity and Germanic customs."[182] And so on. In a word: the pattern is very old in Europe.

The matter of dates, or age, is certainly crucial. Western Europe was undergoing major transformations from the seventeenth century onward, and a delayed and lower marriage rate could well be explained by a number of the newer facts: mobility, loss of holdings due to enclosures, urbanization, and, later, the well-understood demographic and social effects of the industrial revolution. Even the sixteenth century was somewhat chaotic in western Europe. But if the marriage pattern emerged before 1492, a date before these disruptions began, then one can speak of a definite "European pattern," not merely a "preindustrial" or "traditional" pattern. And one can begin to build a general causal theory for the subsequent changes—the "European miracle."

Additional propositions were added. It had long been known that nuclear family households and neolocal residence (the married couple establishes a household well separated from those of the parents) have been characteristic of western Europe during recent centuries, and this is to be expected as a feature of modernization. It fits with standard modernization theory: the process is supposed to lead from extended family to nuclear family. But the European miracle historians now argue that nuclear families and neolocal residence are part of a unique "European familial system" (Laslett).[183] And again they locate its origin far back in history. And again they sweepingly assert that non-Europeans lack these patterns. In fact, there is good evidence that nuclear-family households were common in many parts of non-Europe. Taeuber, analyzing Buck's data, found that more than 60% of peasant households in China were nuclear in the early twentieth century.[184] These are not historical data, but the point is that China is supposed, according to modernization theory, to be especially "traditional" as regards family and many other things; indeed, there is common confusion between the notion of the Chinese lineage or "sib"—made famous by Weber—and the idea of the extended family. Larger, extended families are uncommon in Latin America. In India, as well as in China, the notion of an extended family is ambiguous because of the ambiguity of the notion of "neolocal residence" (in a cramped village, house-building space can be a problem), and the association sometimes found between neolocality and available land upon which to begin a small holding. Further confusion is added over the association between extended (and/or joint) family and inheritance rights, power, mobility, and more.[185] Meanwhile, back in Europe, the model of a supposedly characteristic nuclear–neolocal pattern is subject to very serious questions for the Middle Ages. (For even earlier times the idea of a unique pattern is speculative, and belongs with other old and suspect ideas—discussed earlier in this chapter—about the ancient Germanic tribes and their unique individualism, progressiveness, and the like.) Medieval marital residence patterns may sometimes have reflected manor rules (such as assignment of holdings) rather than cultural residence rules in circumstances of serfdom and insecure tenancy, and a distinction has to be made between different types of rural regions, such as the areas with off-farm employment (like woolen areas of southeastern England), those with "frontier" characteristics, and so on. There is even reason to question the notion of a persistent west European marriage pattern.[186] But even granting the generalization about western Europe, there is no basis for considering the European pattern to have been, historically, unique.[187]

Finally, the model is embellished with some rather gaudy ornamentation. It is claimed that the unique European pattern is not merely a matter of neolocality, nuclear households, and age and frequency of marriage. West European marriages were grounded in *love*. Elsewhere, marriages were arranged. (But arranged marriages seem to have been the rule in premodern Europe, prior to the periods of high mobility and social disruption. The notion, regularly invoked here, that non-European couples are strangers to one another prior to their being shoved into marriage is colonial-era prejudice supported by a few unusual cultural situations. Romantic love in this argument is considered an attribute of European rationality. Non-European couples are certainly as loving as European couples.) It is claimed, next, that the unique European family, because it is nuclear, produces a unique European personality type. The theory is that the small (European) household leads inevitably to individualistic, competitive, acquisitive, yet caring behavior. Mann and Jones, as we saw earlier, put forward the image of an ancient peasant household, a Hansel and Gretel house deep in the forest, as the historical source of this individualistic pattern. Macfarlane, who presents perhaps the most extreme form of this theory (and is chided by some of his colleagues not for his theory but for his claim that it applies mainly to the *English*), argues that the early medieval English family produced a person with the psychological and behavioral traits of (in essence) Weber's capitalist personality, that the causal chain runs from tribal customs and religion to family, from family to personality, and from personality to the beginnings of capitalism (a point we return to below). This argument is advanced mainly by erecting a model of what Macfarlane considers to be "peasant society," with a peasant family and peasant mentality. Early English rural folk don't fit this model. Therefore they were not peasants, and did not have the traditional traits of peasantry, like a traditional family type and traditional mentality.[188] As many have commented, Macfarlane's notion of "peasant" is a straw man, uncharacteristic of modern non-Europe, much less of historical non-Europe, and his view of the medieval English farm folk as nonpeasants is historically invalid.[189]

The theory of the unique European family is an important part of the European miracle theory as it is being propagated today. It is used in two distinct arguments. The first of these combines the family theory with Malthusianism. The argument runs as follows: Humans in general are not rational enough to control their sexual behavior and limit the number of offspring to suit the prevailing conditions of food supply and the like. Therefore, ordinary humans experience permanent, or periodic, crises, in which overpopulation leads to famine and war and pestilence, after which the now-reduced population begins again to overproduce offspring. The root axiom is *irrationality*. People behave not intelligently, but (as Malthus said 200 years ago) rather like barnyard beasts.[190] The historical result of this process is to *prevent development*. Any improvement in, for instance, productive technology merely leads to population growth, to crisis, to depopulation, and thus back to the status quo ante—a cycle of stagnation. This is one and perhaps the crucial cause of the nondevelopment of non-European societies in general. Europeans, by contrast, have always (or perhaps for just a millennium or so) exhibited *rational* behavior in all matters relating to population. This includes decisions about marriage and about childbearing. The unique European late-marrying, nuclear, neolocal, companionate family is the crucial institution in which this rational decision-making process takes place. Thus the European family has permitted Europeans (or west Europeans, or northwest Europeans) to check population growth, and thereby accumulate material wealth that would otherwise be dissipated in the feeding of excess babies. This primordial accumulation underlies the permanent progress of Europe. This theory, in one form or another, is advanced by many of the historians whom we have been discussing, among them Crone, Hall, Jones, Laslett, Macfarlane, and Mann.[191] Hall expresses one form of the theory in the following comment (part of which we quoted earlier):

> [The] European family has long been small, late marrying, nuclear and notably sensitive to Malthusian pressure. . . . The expansion of the European economy did not occur *laterally*, as in late traditional China, because improvements in output were not eaten up by a massive growth in

population. The ratio between population and acreage in Europe remained favourable ultimately because of the relative continence of the European family.[192]

The Malthusian theory reduces humans to beasts. But even for those scholars who accept Malthusianism, the notion that the European family staves off the Malthusian disasters is hardly credible.

The second way in which the "unique European family" theory is used in arguments for the European "miracle" is a matter of deducing ways that the European family produces various sorts of uniquely progressive attitudes and actions on the part of Europeans, attitudes and actions that then lead toward the miracle. A very widely used form of this argument is expressed thusly by Laslett:

> [The] European familial system may have been responsible for a whole series of features conducive to economic progress, and perhaps innovation. The conditions laid down for marriage and procreation imposed on all individuals . . . the necessity of saving, accumulating. . . . [The] European familial system fostered the spirit of hoarding and parsimony.[193]

The family thus generates the personality traits of an incipient capitalist, and does so deep in the Middle Ages. This, too, is hardly credible and hardly in accord with the realities of life (and psychology) before 1492. Laslett, like Macfarlane and many others, makes the false assumption that a typical medieval European family does, in fact, have choices permitting saving, hoarding, accumulating, and so on. This argument requires that the peasant family be a substantial landholder, or at least a tenant with firm security, such that capital accumulation can take place and not be bled off by master or landlord. There is much dispute about the degree of landownership in late-medieval western Europe, but it is certain that family behavior of the sort described by Laslett could not have been common under serfdom and, after the decline of serfdom, could not have been common in places where peasants did not own the land. And this seems to have been the rule, not the exception.[194] It appears that this argument is, again, a telescoping of history. The entrepreneurial yeoman farmer of, say, the eighteenth century, is pushed back into the Middle Ages, and an effect becomes a cause.

Many other European-family-to-European-miracle arguments are common in present-day scholarly discourse, but space permits me to give only one additional example, from Lawrence Stone:

> There are several consequences which must have followed from the late marriage pattern. . . . [It] is reasonable to assume that for many young men

this delay involved considerable sexual denial at a time of optimum male sexual drive. . . . If one follows Freudian theory, this could lead to neuroses. . . . [It] could help to explain the high level of group aggression, which lay behind the extraordinary expansionist violence of western nation–states at this time. It could also have been a stimulus to capitalist economic enterprise . . . stimulating saving in order to marry, and generating activist social and economic dynamism.[195]

Again:

The sublimation of sex among young male adults may well account for the extraordinary military aggressiveness, the thrift, the passion for hard work, and the entrepreneurial and intellectual enterprise of modern Western man.[196]

Europeans conquered the world because their young men were sexually frustrated.

Space does not permit us to list and discuss all of the other common or garden-variety explanations for the so-called European miracle. Some additional explanations will be mentioned in various contexts at later points in this book. But I hope that the discussion in this chapter has sufficiently made the point that *no* characteristic of Europe's environment, Europe's people, or Europe's culture, at any time prior to 1492, can be convincingly shown to have had anything to do with the fact that Europe developed while other civilizations did not do so.

I will try, in the next two chapters, to show that the whole question must be phrased in a different way and answered in a different way. Europe did *not* rise, relative to the other civilizations, prior to 1492. When Europe *did* rise, after 1492, this reflected, not some special quality of "Europeanness," but the immense wealth which came into Europe as a result of colonialism in the sixteenth century and thereafter.

NOTES

1. Although Eric L. Jones's 1981 book *The European Miracle* popularized the phrase "the European miracle," it had been in use for a long time with the same essential meaning: the unique rise of Europe before or during the Middle Ages. Not every historian would describe this as a "miracle," but the term has wide acceptance, as evidence the fact that an international symposium entitled "The European Miracle" was held at Cambridge University in 1985.

2. Apparently my 1976 article, "Where Was Capitalism Born?" was the first publication to reject the "miracle" theory absolutely, that is, without qualification. In 1990 Samir Amin, commenting on a later paper of mine, indicated his basic

agreement with this rejectionist position. (Amin, "Colonialism and the Rise of Capitalism: A Comment," 1990; Blaut, "Colonialism and the Rise of Capitalism," 1989; also see Blaut, "Fourteen Ninety-Two," 1992; and Amin, "On Jim Blaut's 'Fourteen Ninety-Two,'" 1992.) A few other historians have taken positions approaching this one; their views are discussed later in this chapter.

3. Another seminal work was Cyril Black's *The Dynamics of Modernization: A Study in Comparative History* (1966). Other such works in various fields will be discussed later in this chapter.

4. Perhaps one reason was the maturation of the discipline of history itself. Another was the influence, not entirely salutary, of positivistic scientific method on history, which led to an attempt to specify variables and "factors," and, if possible, quantify each of them. Another reason was the loss of faith in the nineteenth-century idea of inevitable progress: after nearly a half century of chaos and war, progress definitely did not appear to be something natural and inevitable; it had to be explained and also produced. Still a third reason was the general secularization of European thought, including history, such that human events could not be assumed to reflect guidance of a higher power. Another reason, possibly a very important one, was the general development of academic disciplines, and their involvement in (and nutrition from) international and domestic policy. This implied that each discipline, seeing the world with some degree of bias in favor of its own subject-as-factor (for economists, the market factor; for psychologists, the motivation factor; for sociologists, the demographic factor and the social–structural factor; for geographers, the resource factor; etc.) tended to argue in favor of historical models that treated *our* factor as crucial and those of *others* as secondary. Since many social scientists wrote history in this process, this led to some special pleading.

5. I do not mean to imply that all of this was a dominant theory for the discipline of history as a whole. The great majority of historical scholars worked on smaller problems, carefully delving into events and developing limited explanations for those events. The modernization perspective influenced some of the smaller generalizations, for instance proffering explanations in which the factors most supportive of modernizationism were favored—factors like population, technology, and the like. And there were some areas of research where lack of attention to the history of non-Europe was a critical source of error (most notably, as we will see, in studies of the history of European technology). In addition, there were (and are) many different points of view in the vast and diverse field of professional history, so it would be questionable to characterize a particular historiographic period as being dominated by a particular governing theory (or "paradigm"). I suspect that my own concern with the body of literature relating to the "European miracle" problem probably leads me to overemphasize the significance of the modernization view on history as a whole. It should also be noticed that most of the prominent writers about the specific problem of explaining the unique rise of Europe were specialist historians—economic historians, historical sociologists, and so forth—and not historians of the standard breed.

6. Cabral, *Unity and Struggle* (1979).

7. As to the present and future, Third World intellectuals tended to make two arguments. Those who supported the idea of a capitalist form of development argued that economic development must consist of the defense of native capital against the corrosive diffusion into one's country of the economic and political dominance of European countries and corporations. For socialists, influence and dominance by international capitalism quite obviously had to be rejected. Both groups tended to

adopt "dependency theory" or "underdevelopment theory," which was a theory both of history and of modern social processes and development. A relatively small minority of Third World intellectuals, usually reflecting the thinking and interests of the very wealthy and very right-wing sectors, welcomed the idea of economic domination by foreign capitalist interests. Since the wealthy social sectors dominated most Third World societies, this minority point of view often determined policy. Also, it received much more prominence than it deserved in the journals of the First World.

8. See James, *A History of Pan-African Revolt* (1938), *The Black Jacobins* (1938), "The Atlantic Slave Trade and Slavery" (1970); Williams, *Capitalism and Slavery* (1944). We discuss these matters in Chapter 4.

9. Amin, *Accumulation on a World Scale* (1974) and later works. My articles "Geographic Models of Imperialism" (1970) and "Where Was Capitalism Born?" (1976) laid out the skeleton of a general theory.

10. I deal with this matter in *The National Question* (1987b).

11. See Van Leur's 1934 essay "On Early Indonesian trade," reprinted in his *Indonesian Trade and Society* (1955).

12. Duyvendak, *Ma Huan Re-examined* (1933); Needham and collaborators, *Science and Civilization in China*, published in 6 volumes between 1965 and 1984; Wheatley, *The Golden Khersonese* (1961) and *The Pivot of the Four Quarters* (1971); Elvin, *The Pattern of the Chinese Past* (1973).

13. Amin, *Unequal Development* (1976), *Eurocentrism* (1988), "Colonialism and the Rise of Capitalism: A Comment" (1990).

14 Bernal, *Black Athena*, vol. 1 (1987) and vol. 2 (1991).

15. Mention should also be made of Eric Wolf's 1982 book *Europe and the Peoples Without History* which provides a useful and important survey of the history of both European and non-European civilizations and shows how unconvincing is the theory that non-European civilizations, historically, were stagnant and unprogressive (that they were "peoples without history"). Wolf, however, stops short of questioning the truly crucial Eurocentric belief that Europeans were *more* progressive than non-Europeans in several ways that are crucial to the "European miracle" theory, and so he does not directly confront that theory. (It should be noted that most mainstream historians no longer argue that non-European civilizations are, or were, totally unprogressive, totally "unhistorical," arguing instead about slow rates of change, "blockages" inhibiting change, and the like—a difference of phrasing which, as we will see, is not always a difference of argument.)

16. It is always risky to try to explain broad changes of fashion in scholarship, especially when the changes are still underway, so this interpretation cannot be more than a hunch, or hypothesis. Scholarly attitudes toward the Third World were very positive in the period of anticolonial and civil rights struggles. After the late 1960s the mood changed. Not only did more conservative views come to dominate the Western world, but rather unexpected difficulties emerged in the Third World itself: national conflicts, failure of development programs, and more. Western scholarship had never really abandoned the traditional view of Europe and its relations to non-Europe, including the diffusionist view of colonialism, and it appears that this older paradigm, never abandoned, simply became, once again, thoroughly dominant. Certainly the attention that had been paid previously to dependency theory and related views fell away among mainstream scholars. Among Marxists the process was even more dramatic, because it was less expected. In brief, Eurocentric Marxists who dismissed the role of the Third World both historically and in the present now became virtually

the only Marxists within the academic world to pronounce upon issues related to the Third World. It became quite fashionable to insist, once again, that only the working class of the advanced capitalist countries can bring about socialism because each stage of history commences in this part of the world (Inside) before it spreads to the rest of the world (Outside). In the conservative camp, not only did Eurocentric views regain their hegemony, but one began, now, to hear whispers of views not far distant from racism: views about Third World peoples not having the potential to develop.

17. See Brenner, "The Origins of Capitalist Development: A Critique of Neo-Smithian Marxism" (1977), "Agrarian class structure and economic development in pre-industrial Europe" (1985), "The Agrarian roots of European capitalism" (1985b); Anderson, *Passages from Antiquity to Feudalism* (1974), *Lineages of the Absolute State* (1974); Warren, *Imperialism: Pioneer of Capitalism* (1980).

18. I will discuss only the more important beliefs, perhaps neglecting a few of these. And I will give just enough evidence to show that these beliefs are not self-evidently true. Much fuller evidence is given, to dispute some of the beliefs, in Chapters 3 and 4.

19. See Chapter 1, note 2.

20. Bowler, *The Invention of Progress* (1989); Harris, *The Rise of Anthropological Theory* (1968); Gossett, *Race: The History of an Idea in America* (1963); Jackson, *Race and Racism: Essays in Social Geography* (1987); Stocking, *Race, Culture, and Evolution* (1968); Trigger, *A History of Archeological Thought* (1989).

21. See Bernal, *Black Athena*, vol. 1 (1987).

22. See Gossett, *Race* (1963).

23. Blaut, "The Theory of Cultural Racism" (1992).

24. "Hereditary neuralgia of the presumably strong tendency toward hysteria and autohypnosis of the Indian . . ." Max Weber, *The Religion of India* (1967), p. 387. This postulate is basic to Weber's analysis of Brahminism as prime cause for India's lack of development.

25. "[The] . . . negroes long ago showed themselves unsuitable for factory work and the operation of machines; they have not seldom sunk into a cataleptic sleep. Here is one case in economic history where tangible racial distinctions are present" Weber, *General Economic History* (1981), p. 379. Although Weber says here that this is merely "one case" of racial influences, note that the case is crucial for Weber's analysis both of "rationality" and of modernization; the "racial distinctions" here separate out Africans in a fundamental way: a clear case of moderate yet crucial racism. In the same vein, "it was found that the American Indians were entirely unsuitable for plantation labor" (p. 299).

26. Weber, *The Religion of China* (1951), pp. 231–232. Weber considers these to be "racial qualities of the Chinese" (p. 230) although cultural factors may be involved.

27. Weber, *The Protestant Ethic and the Spirit of Capitalism* (1958), p. 30. Weber says here that "it would be natural to suspect that the most important reason" for the rationality of the Occident lies in "differences of heredity," and "the importance of biological heredity," he thinks, is "very great" (p. 30). But we do not yet know how to measure its influence, so our attention should focus mainly on the cultural factors (pp. 30–31). This was typical moderate racism (as I call it) for the time the work was published (1904–1905).

28. I have not discussed the question whether biological racism remains important today as a genuinely implicit theory, that is, one still accepted but not consciously so. I suspect that it does. Some Eurocentric historians hold to positions regarding individual differences between Europeans and non-Europeans that are so

extreme, and so bigoted, that it is difficult to avoid the conclusion that at least a few of them may hold to an implicitly racist infratheory, perhaps unconsciously believing, that is, that the inferiority of non-Europeans is genetically determined.

29. Among many sources on the racism fallacy, see Franz Boas' classic book, *Race, Language, and Culture* (1948); also Blum, *Pseudoscience and Mental Ability* (1978); Gossett, *Race* (1965); Haller, *Outcasts from Evolution* (1971); Jackson, *Race and Racism* (1987); Magubane, *The Ties that Bind: African-American Consciousness and Africa* (1987); Gill and Levidow, *Anti-Racist Science Teaching* (1987).

30. The idea that demographic behavior is not fully under social control, is a primordial biological fact or factor, seems to underlie the thinking of most Eurocentric historians (conspicuously Eric L. Jones, Michael Mann, and John Hall). The basic proposition seems to be that population will necessarily grow beyond the rational limits, and so overpopulation *must* result unless societies find nondemographic solutions, notably by increasing food supply to feed the inexorably increasing population. See, for instance, Michael Mann, *The Sources of Social Power*, vol. 1 (1986): If agricultural yields had not increased in medieval Europe, "the continent would . . . have experienced a similar Malthusian cycle every century or so—and would not have generated capitalism" (p. 402). This view is even encountered, occasionally, in the radical and feminist movements. Note the following view put forward by a Marxist–feminist: "[We] assume (given the ubiquity of the sex drive . . .) that the total number of conceptions in a population will tend to exceed those desired within a given incentive structure, and this unintended surplus will be greater, the more imperfect are the means at hand. . . . A given fertility pattern then is taken to be 'rational' . . . *except* for a small but variable excess [of births]," Seccombe, "Marxism and Demography," (1983), p. 31. The context is a discussion of European medieval social evolution, but the argument is given mainly in support of Brenner's theory about the strictly European rise of capitalism. Most radicals and feminists reject Malthusianism and would reject Seccombe's view as Malthusian.

31. Jones, *The European Miracle* (1981), p. 3. Part of the passage is a quotation.

32. Jones, *The European Miracle* (1981), p. 219.

33. Hall, *Powers and Liberties: The Causes and Consequences of the Rise of the West* (1985), p. 131. Italics added.

34. As F. Hassan points out, "Among human populations the practice of population control in one form or another is universal." ("Demographic Archeology," 1978, p. 71.)

35. Some families will of course have more conceptions than desired, and others fewer, because birth control techniques are imperfect, but the average for the larger group will be broadly in line with the group's values and population targets. The techniques run the gamut from variable age of marriage and variable dowry or bride price, to complexity of marriage rules (defining who, in a kin system, is eligible to be one's spouse), to timing of sexual relations, to the use of antifertility and abortive devices, to infanticide, and other practices.

36. Much of the scholarship comes from India, where colonial ideology used to claim that poverty is a result of people having too many children. Now demographers and other social scientists have disproven this myth. See, for instance, Mamdani, *The Myth of Population Control* (1972), and Nag, "How Modernization Can Also Increase Fertility" (1980). For Africa, see, for example, Kitching, "Proto-Industrialization and Demographic Change" (1983); Swindell, "Domestic Production, Labor Mobility, and Population Change in West Africa, 1900–1980" (1981); Cordell and Gregory, the introduction to *African Population and Capitalism* (1987).

37. See, for instance, numerous studies showing the plasticity of birthrates, including Nag "How Modernization Can Also Increase Fertility" (1980), Collyer, *Birth Rates in Latin America* (1965), and Harewood, "Population Growth in Grenada" (1966).

38. See Aston and Philpin, eds., *The Brenner Debate* (1985), especially the introduction by R. Hilton.

39. Perhaps more readily, because the Malthusian explanation for poverty in Third World countries has been a rather troublesome issue for scholars and planners in these countries, and it has been important to show that poverty in their countries is not somehow caused by the demographic misbehavior of ordinary people.

40. For instance Montesquieu, in *The Spirit of the Laws* ([1748] 1949) "People are . . . more vigorous in cold climates" (pt. xiv.2). "There are countries where the excess of heat enervates the body, and renders men . . . slothful and dispirited" (pt. xv.7).

41. "Africa" almost always refers to "Africa south of the Sahara" in the discourse I am criticizing, so I will use "Africa" in the same sense in the present discussion.

42. Blaut, "The Ecology of Tropical Farming Systems" (1963).

43. See, for example, Collins and Roberts, eds., *Capacity for Work in the Tropics* (1988), which fails to find convincing evidence suggesting negative tropical effects.

44. The two citations (Gilfillan, "The Coldward Course of Progress," 1920, and Lambert, "The Role of Climate in the Economic Development of Nations," 1971) are hardly indicative of modern scholarly literature.

45. India's "debilitating climate" is one reason why India fell behind Europe, according to Jones, *The European Miracle* (1981), p. 198.

46. Among the pioneer works which have importance for peasant agriculture, we may mention Fred Hardy's "Some Aspects of Tropical Soils" (1936) and various of his articles in *Tropical Agriculture*, and the work of Robert Pendleton (especially Pendleton, "Land Use in North-Eastern Thailand," 1943, and Prescott and Pendleton, *Laterites and Lateritic Soils*, 1952), and G. Milne "A Soil Reconnaissance Journey Through Parts of Tanganyika Territory" (1947). The first comprehensive textbook on tropical soils embodying modern knowledge is Mohr and van Baren, *Tropical Soils* (1954).

47. See Nye and Greenland, *The Soil Under Shifting Cultivation* (1960); Blaut, "The Nature and Effects of Shifting Agriculture" (1962); Ahn, *West African Soils* (1970).

48. See Miller, *Way of Death* (1988); Curtin, *The Rise and Fall of the Plantation Complex* (1990). The view is echoed in many world history textbooks, for example, Roberts, *The Hutchinson History of The World* (1987), pp. 54–56; McNeill, *A World History* (1967), pp. 273–278.

49. See Wilken, *Good Farmers: Traditional Agricultural Resource Management in Mexico and Central America* (1987); also Nye and Greenland, *The Soil Under Shifting Agriculture* (1960); Blaut, "The Ecology of Tropical Farming Systems" (1963).

50. Denevan, *The Native Population of the Americas in 1492* (1976).

51. See Cockburn and Hecht, *The Fate of the Forest* (1989).

52. These marginal areas often are regions in which agriculture is practiced, in preference to some less intensive system of land use, because human communities have been pushed off better land by historical forces, notably colonialism.

53. For a typical instance, the Oxford historian J. M. Roberts, in his popular world history textbook, *The Hutchinson History of the World* (1987), makes the following quite ignorant statement: "Probably the greatest importance of [the] spread of iron-working [in tropical Africa] was the difference it made to agriculture. It made possible a new penetration of the forests, new tilling of the soil (which may be

connected with the arrival of new food-crops from Asia . . .) . . . This suggests once again the important limiting factor of the African environment. Most of the continent's history is the story of response to influences from the outside [including iron-working and new crops]" (pp. 511–512). African farmers, like farmers in Europe and many other regions, practiced agriculture with stone tools before iron arrived, and continued to do so afterward whenever and wherever iron was hard to obtain. On the age of ironworking in tropical Africa, see, for example, Wai-Andah, "West Africa Before the Seventh Century" (1981) and Sinclair, "Archeology in Eastern Africa" (1991).

54. See, for example, Roberts (note 53 above); also Irwin, "Sub-Saharan Africa," in Garraty and Gay, eds. The Columbia History of the World (1981), p. 299.

55. See for instance Irvine's classical work, A Textbook of West African Agriculture (1934); Coursey, Yams (1967).

56. According to Jones, The European Miracle (1981), "the Negroid peoples . . . were still pushing east and south into the territories of the pygmies and the bushmen when the Boers undertook the Great Trek north from the Cape in the 1830s," (p. 155). Also see Roberts The Hutchinson History of the World (1987), p. 178.

57. Curtin, Economic Change in Pre-Colonial Africa (1975); Curtin, The Rise and Fall of the Plantation Complex (1990); Miller, Way of Death (1988).

58. See Wisner and Mbithi, "Drought in Eastern Kenya" (1974); Wisner, Power and Need in Africa (1989); O'Keefe and Wisner, "African Drought: The State of the Game" (1975).

59. The same arguments apply, with no need for qualification, in Western Hemisphere agriculture. In relatively limited parts of the South American rain forest the ecosystem is so fragile, as a result of local geological conditions (rocks forming infertile kaolinitic clay soils in some areas, almost pure sands in other areas), that cropping systems must employ shifting agriculture (or tree crops). But such areas are the exception within the present-day distribution of rain forest environments. In general, the great sweep of Amazonian and Guianan rain forest is a zone of nonsedentary agriculture because of cultural–historical factors, notably depopulation and the massive increase in cattle ranching. It is in fact cattle ranching, and not shifting agriculture, that leads to agriculturally caused long-term environmental degradation in the rainforest region, because (1) ranchers burn forest to the maximum extent possible and without control, in order to expand the area of pasture, and (2) the resulting pasture leads to steady soil degradation because coarse pasture grasses do not maintain the soil ecosystem as does the original forest. Shifting cultivators, by contrast, burn only under controlled conditions, burn small areas only, and carefully encourage the regrowth of forest. If the forest disappears, their livelihood disappears. Shifting agriculturists should not be blamed for deforestation anywhere in the humid tropics. See Hecht and Cockburn, The Fate of the Forest (1989); Blaut, "The Nature and Effects of Shifting Agriculture" (1963).

60. Buckle, The History of Civilization in England, 2nd ed., (1913), chap. 2. Also see Bowler, The Invention of Progress (1989), pp. 28–29.

61. Marx, Capital, vol. 1 (1976), p. 513n.

62. Jones in fact manages to use both of the opposing theories toward the same end. In wetter regions of Africa "living was easy." In drier areas "agriculture was not productive." Jones, The European Miracle (1981), p. 154. Also see Jones, Growth Recurring: Economic Change in World History (1988).

63. "In Africa the bountifulness and extent of the land makes for a mobile peasantry, necessarily therefore poor material on which to build states. Something like

this is probably true of all slash-and-burn agriculture," John Hall, *Powers and Liberties* (1985), p. 27. ("Slash-and-burn agriculture" is shifting agriculture.)

64. Laibman, "Modes of Production and Theories of Transition" (1984), p. 284. However, Laibman's overall argument is not at all Eurocentric.

65. Buckle, *History of Civilization in England*, vol. 1, 2nd ed. (1913), p. 93: "[The] great plagues by which Europe has at different periods been scourged, have, for the most part, proceeded from the East, which is their natural birthplace, and where they are most fatal. Indeed, of those cruel diseases now existing in Europe, scarcely one is indigenous; and the worst of them were imported from tropical countries in and after the first century of the Christian era."

66. The theory that the HIV virus which causes AIDS in humans is another one of these African plagues descending upon the Western World may very well be just the newest myth in this old diffusionist tradition. Whether or not this virus originated in Africa, which has not been proven, the myth has already taken on ugly overtones, as in the completely unfounded (yet classical) belief that AIDS was transmitted from monkeys to humans because of some strange sexual practices in some obscure African tribes. (This myth is naively reported in Shannon and Pyle, "The Origin and Diffusion of AIDS," 1989. See the critique of this paper by Watts and Okello, "Medical Geography and AIDS," 1990. Also see R. C. and R. J. Chirimuuta, *AIDS, Africa and racism*, 2nd ed., 1989.)

67. McNeill, *Plagues and Peoples* (1976), p. 43.

68. See, for instance, Giblin, "Trypanosomiasis Control in African History: An Evaded Issue?" (1990); Turshen, "Population Growth and the Deterioration of Health: Mainland Tanzania, 1920–1960" (1987).

69. Wittfogel, *Oriental Despotism* (1957).

70. See Venturi "The History of the Concept of 'Oriental Despotism' in Europe" (1963); P. Anderson, *Lineages of the Absolute State* (1974); B. Chandra, "Karl Marx, His Theories of Asian Societies, and Colonial Rule" (1981).

71. Similar judgments are still commonly made by theoreticians of the European miracle. For instance, John Hall, *Powers and Liberties* (1985), p. 12: "Chinese society was stuck in the same stage for over two thousand years, while Europe, in comparison, progressed like a champion hurdler."

72. And also earlier. It is discussed for instance by Montesquieu, Bernier, Adam Smith, and Hegel (see, for instance, "Introduction," and "The Oriental World" in Hegel's *Philosophy of History*, 1956). Also see the historical reviews in Venturi, "The History of the Concept of 'Oriental Despotism' in Europe" (1963); P. Anderson, *Lineages of the Absolute State* (1974); and B. Chandra, "Karl Marx, his Theories of Asian Societies, and Colonial Review" (1981).

73. This idea, too, had its forerunners. Possibly Marx got the idea from Karl Ritter, his professor of geography at Berlin.

74. Marx and Engels, *Selected Correspondence* (1975).

75. Marx and Engels considered other factors as well, and it is fair to say that their analysis remained speculative. I believe that Engels withdrew from the idea of "Oriental despotism" in late life. See the discussion of this topic in P. Anderson, *Lineages of the Absolute State* (1974); Blaut, "Colonialism and the Rise of Capitalism" (1989); and B. Chandra, "Karl Marx, His Theories of Asian Societies, and Colonial Rule" (1981).

76. See Laibman, "Modes of Production and Theories of Transition" (1984), and Bailey and Llobera, *The Asiatic Mode of Production* (1981).

77. Weber, *The Agrarian Sociology of Ancient Civilizations* (1976), pp. 84, 131, 157, *The Religion of China* (1951), pp. 16, 21, 25, and "The Origin of Seigneurial Proprietorship," part 1, chap. 3, esp. pp. 56–57 in *General Economic History* (1981). Also see McNeill, *Plagues and Peoples* (1976), pp. 93, 207, 259.

78. Weber, *The Agrarian Sociology of Ancient Civilizations* (1976), pp. 157–158. Also, on p. 84: "The basis of the economy [in Egypt] was irrigation, for this was the crucial factor in all exploitation of land resources. Every new settlement demanded construction of a canal ... Now canal construction is necessarily a large-scale operation, demanding some sort of collective social organization; it is very different from the relatively individualistic activity of clearing virgin forest. Here then is the fundamental economic cause for the overwhelmingly dominant position of the monarchy in Mesopotamia [and] Egypt."

79. See, for instance, Denevan, *The Aboriginal Cultural Geography of the Llanos de Mojos of Bolivia* (1966), and "Hydraulic Agriculture in the American Tropics" (1982), on ancient drained-field or raised-field agriculture in the tropics; Golson, "No Room at the Top: Agricultural Intensification in the New Guinea Highlands" (1977), on ancient drainage in highland New Guinea; Doolittle, *Canal Irrigation in Prehistoric Mexico* (1990), on drainage and early agriculture in Mexico; Harrison and Turner *Pre-Hispanic Maya Agriculture* (1978), on early drainage in the Maya lowlands.

80. Jones, *The European Miracle* (1981), pp. 8–9.

81. Hall, *Powers and Liberties* (1985), pp. 12–13, 27–28, 36, 42–3, 53, 59, 99, 102, 137.

82. Hall, *Powers and Liberties* (1985), p. 11. Also see pp. 41, 123, 132.

83. Mann, *The Sources of Social Power* (1986). Also see his essay, "European Development: Approaching a Historical Explanation" (1988).

84. Mann, *The Sources of Social Power* (1986), p. 94.

85. Mann, *The Sources of Social Power* (1986), p. 179.

86. The fact that Mann attributes the ancient European takeoff mainly to chariot warfare and iron-plow rainfall-farming, yet concedes that both innovations were initiated by Middle Easterners themselves, suggests to me that Mann's fundamental causal reasoning centers on the notion of European rationality: regardless of who invented these things, the Europeans put the things to work. That this Weberian notion is indeed basic for Mann will be demonstrated later in this chapter.

87. Bray, *Agriculture*, vol. 6, part 2, of Needham and collaborators, *Science and Civilization in China* (1984).

88. Blaut, "Two Views of Diffusion" (1977).

89. Mann, *The Sources of Social Power* (1986), pp. 247, 406, 408, 412, 504–510, 520, 530, 539–540.

90. Mann, *The Sources of Social Power* (1986), p. 509.

91. Hall, *Powers and Liberties* (1985), p. 99.

92. Hall, *Powers and Liberties* (1985), p. 110.

93. Jones, *The European Miracle* (1981), p. 10.

94. Jones, *The European Miracle* (1981), p. 47.

95. Jones, *The European Miracle* (1981), p. 8. Jones makes the common error of assuming that productivity per worker is low in irrigated agriculture in comparison with unirrigated agriculture. This is not the case. Even with draft animals, medieval peasant production per worker was not high. And draft animals are used in irrigated farming, abundantly so in some Asian wet-rice farming systems.

96. Mann, "European Development" (1988), p. 10, *The Sources of Social Power* (1986), p. 406; Jones, *The European Miracle* (1981), pp. 90, 227; Crone, *Pre-Industrial Societies* (1989), p. 150; McNeill, *Plagues and Peoples* (1976), p. 295.

97. Mann, *The Sources of Social Power* (1986, chap. 5). Mann in fact builds a theory of ancient warfare upon the basis of this sort of calculation, here also ignoring the fact that armies then as now provision themselves and their animals on the route of march.

98. Jones, *The European Miracle* (1981), chap. 2 and elsewhere. Hall, *Powers and Liberties* (1985), p. 132, makes the same claim, citing Jones.

99. Hall, *Powers and Liberties* (1985), p. 111; Jones *The European Miracle* (1981), pp. 90, 105, 107, 226–227; Mann, "European Development: Approaching a Historical Explanation" (1988), p. 10; Mann, *The Sources of Social Power* (1986), p. 406.

100. Occasionally "the criminal mind" was seen to reflect another dimension of contrast.

101. Lévy-Bruhl, *How Natives Think* (1966).

102. See, for instance, Boas, *The Mind of Primitive Man* (1938); Radin, *Primitive Man as Philosopher* (1927); M. Mead, *Growing Up in New Guinea* (1930).

103. Stocking, *Victorian Anthropology* (1987), p. 59; Bowler, *The Invention of Progress* (1989), p. 66; Whitman "From Philology to Anthropology in Mid-Nineteenth-Century Germany" (1984); Bernal, *Black Athena*, vol. 1 (1987); Said, *Orientalism* (1978). A recent expression of this theory came in a Soviet debate about "Oriental despotism" (Lelekov, "Round-Table: State and Law in the Ancient Orient," 1978). L. Lelekov, a historian, claimed that words meaning "freedom" and "right" were basic in the original Indo-European language or languages but not in Near Eastern languages, and asserted that this must have affected "social thinking" (p. 190). This contention was refuted by the philologist V. Ivanov (p. 193).

104. See Dalal, "The Racism of Jung" (1988). In Jung's work, see in particular *Psychological Types* (1971) (for instance: "[If] we go right back to primitive psychology, we find absolutely no trace of the concept of the individual," p. 10); *Memories, Dreams, Reflections* (1963), and "The Dreamlike World of India," in *Civilization in Transition* (1927). See also the 1954 work by Jung's disciple Erich Neumann, *The Origins and History of Consciousness* (for instance: "The evolution of consciousness as a form of creative evolution is the peculiar achievement of Western man . . . The creative character of consciousness is a central feature of the cultural canon of the West . . . In stationary cultures, or in primitive societies where the original features of human culture are still preserved, the earliest stages of man's psychology predominate," pp. xviii–xix).

105. Piaget, *Psychology and Epistemology* (1971), p. 61.

106. See, for example, Werner and Kaplan, *Symbol Formation* (1964).

107. See the first 16 volumes (through 1985) of *The Journal of Cross-Cultural Psychology* for a great many examples of this phenomenon. For this period something like one-tenth of all the empirical articles in this U. S. journal were studies by white Southern Africans purporting to show the cognitive inferiority of black Africans.

108. Rogers, *The Diffusion of Innovations* (1962); Rogers and Shoemaker, *Communication of Innovations* (1971), pp. 187–191; McClelland, *The Achieving Society* (1961); Hagen, *On the Theory of Social Change* (1962) and "A Framework for Analyzing Economic and Political Change," in Brookings Institution, ed., *Develop-*

ment of the Emerging Countries: An Agenda for Research (1962), pp. 1–39. I have cited only the initial statements: much literature emerged from these works.

109. S. Marglin, "Losing Touch: The Cultural Conditions of Worker Accommodation and Resistance," in F. and S. Marglin, eds., *Dominating knowledge: Development, Culture, and Resistance* (1990).

110. Sack, *Conceptions of Space in Social Thought* (1980). Geographers prominent in the diffusion paradigm discussed here are L. Brown (*The Diffusion of Innovations*, 1981) and P. Gould (*Spatial Diffusion*, 1969). On peasant traditionalism in relation to natural hazards, see G. White, ed., *Natural Hazards* (1974), as an example of the abundant literature. I criticize these and other arguments about non-Western nonrationality in Blaut, "Two Views of Diffusion" (1977), "Diffusionism: A Unitarian Critique" (1987a), and "Natural Mapping" (1991).

111. In the field of education in the United States the dominant standardized college entrance tests (SAT and ACT) are skewed by culture-specific and sex-specific internal (as well as situational) characteristics, such that women do more poorly than men although they actually perform at higher levels in terms of university grades, while the ACT and SAT test scores of Latinos (African-Americans have not yet been studied in this way), which are very low, have no correlation whatever with their performance in college. Thus the tests effectively minimize female and minority college attendance. Why the tests are almost universally used, even so, is a fascinating question. The same biases are so prominent in IQ testing that such tests have been barred as diagnostic instruments in California schools. The "primitive mind" and "primitive languages" biases sometimes combine, as in U.S. debates about the so-called "English-only" issue and about the question whether non-European literature deserves to be included in the college curriculum along with European literature. In Boston not long ago, 30% of Hispanic children aged six to eight were not attending elementary school because their inability to speak English had been judged to be evidence of mental retardation, and Boston claimed not to have sufficient resources to educate these children in "special schools." Testing in general in U. S. education remains very racist.

112. Rorty, *Philosophy and the Mirror of Nature* (1980), and earlier works by Dewey (for instance *The Quest for Certainty*, 1929), Whitehead (for instance, *Modes of Thought*, 1938), and G. H. Mead (for instance, *Philosophy of the Act*, 1938).

113. This is a general assessment. Some anthropologists continue to maintain either the "primitive mind" doctrine or the closely related "traditional mind" doctrine. An example of the latter is George Foster's well-known and influential book *Traditional Cultures* (1962), of the former, Hallpike's *The Foundations of Primitive Thought* (1979). For a critique, see Schweder, "Cultural Psychology: What Is It?" (1990).

114. On the question of Weber's use of "rationality," its basal status in his theorizing, and yet its uncertain definition and provenance, see, for instance: Cohen's introduction to the 1981 ed. of Weber's *General Economic History* (1981), pp. xxv–xxvii; Löwith, *Max Weber and Karl Marx* (1982), pp. 40–42, 53–54, n. 49; Freund, *The Sociology of Max Weber* (1968), pp. 140–149. In Weber, see *General Economic History* (1981), chaps. 29, 30, *The Protestant Ethic and the Spirit of Capitalism* (1958), pp. 13–31, 59–60, 79, 118–120, 191, n. 19, 265, n. 31, *The Religion of China* (1951), chap. 8, *The Religion of India* (1967), p. 387, and other works.

115. Weber, *General Economic History* (1981), p. 161. Also see pp. 339, 355–368. Weber makes a large number of comments about the irrationality of Asians.

116. Weber, *Economy and Society*, vol. 2 (1968), pp. 1212–1374.

117. Numbers in parentheses are page numbers in *The European Miracle*.

118. The ecologists' term "commensalism" refers to a form of tight interspecific cooperation between animal species, but hardly ever is it applied to humans.

119. Typical of Jones's method is to find some objectionable feature of very early Asian society, compare it with some pleasing feature of modern, post-industrial–revolution European society, and then treat *both* features as though they were *permanent* characteristics of the respective societies, giving a picture of awful primitiveness to Asians and marvelous modernity to Europeans thereby.

120. "Frankish feudalism, in many ways proto-typical of later feudalism, was . . . a mixture of the very, very old, deep-rooted drift of 'European' peasant society and of the brand new the opportunistic, the 'un-European,' " Mann, "European Development" (1988), p. 16.

121. Mann, "European Development" (1988), p. 17. Also see his *The Sources of Social Power* (1986), for instance pp. 190, 195, 213, 377, 404, 412, 510. ("At the end of all these processes stood one organic, medium-sized, wet-soil island state, perfectly situated for take-off: Great Britain," p. 510.)

122. Mann, "European Development" (1988), pp. 8–9, 11–12, 15–18 and *The Sources of Social Power* (1986), pp. 377–378, 397–398, 402–408, 412, 500–510. Also, on Europe's rationality and its historical significance, see McNeill, *Plagues and Peoples* (1976), pp. 41, 97, 106–107, 236, 238, 249, 256, 259, 264.

123. See P. Anderson, *Passages from Antiquity to Feudalism* (1974), part 3; Finley, *The Use and Abuse of History* (1975), chap. 6; Aston and Philpin, *The Brenner Debate* (1985), pp. 32–33, 42–51, 59, 63n, 214–215, 234–236, 306–316.

124. White, *Medieval Technology and Social Change* (1962), p. 38; also McNeill, *Plagues and Peoples* (1976), p. 234.

125. White, *Medieval Technology and Social Change* (1962), p. 54.

126. Kosambi, *Ancient India* (1969), p. 89; R. S. Sharma, *Light on Early Indian Society and Economy* (1966), p. 57.

127. White, *Medieval Technology and Social Change* (1962), p. 44.

128. White, *Medieval Technology and Social Change* (1962), p. 56; McNeill, *Plagues and Peoples* (1976), p. 237.

129. See C. T. Smith, *An Historical Geography of Western Europe* (1967), p. 203; Darby, *The Domesday Geography of Eastern England* (1952).

130. Orwin and Orwin, *The Open Fields* (1967), chap. 3; C. T. Smith, *An Historical Geography of Western Europe* (1967), chap. 4.

131. White, *Medieval Technology and Social Change* (1962), p. 57.

132. White, *Medieval Technology and Social Change* (1962), p. 67.

133. White, *Medieval Technology and Social Change* (1962), p. 68.

134. Bray, *Science and Civilization in China*, vol. 6, part 2, *Agriculture* (1984), pp. 304–328.

135. White, *Medieval Technology and Social Change* (1962), p. 76.

136. White, *Machina Ex Deo* (1968). See in particular the essay—quite a famous one—called "The Historical Roots of Our Ecological Crisis" (chap. 5).

137. White, "The Historical Roots of Our Ecological Crisis." In White, *Machina Ex Deo* (1967), p. 85.

138. White, *Machina Ex Deo* (1967), p. 90.

139. Needham et al., *Science and Civilization in China* (1954–1984).

140. Some historians today simply ignore this evidence and repeat the old notions about China's lack of technological prowess. See, for example, Roberts, *The Hutchinson History of the World* (1987), pp. 493–495, 502.

141. The terms "Middle Ages" and "medieval" are conventionally used for most or all parts of the Eastern Hemisphere.

142. Needham et al., *Science and Civilization in China*, vol. 4, part 2 (1965), chap. 27.

143. Needham et al., *Science and Civilization in China* vol. 4, part 2 (1965), p. 33). Arnold Pacey places the innovation in Korea. See Pacey's *Technology in World History* (1990), p. 56.

144. See, for example, Lopez, "Hard Times and Investment in Culture" (1953), Thorndyke, "Renaissance or Prenaissance?" (1943).

145. Cipolla, *Guns, Sails, Empires* (1965), p. 106.

146. Cipolla, *Guns, Sails, Empires* (1965), pp. 108–109.

147. Jones, *The European Miracle* (1981), p. 124.

148. Among the modern states that do not surround such simple ecological core areas: Spain, Italy, Germany, Poland, Greece, Sweden, Russia, etc. Prior to the 20th century, perhaps only parts of Great Britain (southern England) and France (the Paris Basin) came close to fitting this highly abstract model, a model that is useful for some purposes but not for the purpose of explaining the political history of the continent. Some of these core areas are nuclei of states; others are not. In modern Southeast Asia, Burma, Thailand, and Cambodia fit the model as well as any European case.

149. In this connection, see Dirks, *The Hollow Crown* (1987).

150. Hall, *Powers and Liberties* (1985), "States and Societies: The Miracle in Historical Perspective" (1988).

151. Mann, *The Sources of Social Power* (1986), "European Development" (1988).

152. This is not the only example of the way arguments centering on the European state have a way of contradicting one another. The Roman state is said, on the one hand, by Hall among others, to have been a crucial innovation, the source of many political features characteristic of, and only of, Europe. On the other hand, the Roman state is dismissed by others (including Mann) as just another "imperial state," like the despotic Oriental states; Europe supposedly innovated politically by *avoiding* the imperial form of state and developing instead a kind of smaller and somehow more democratic state.

153. Baechler, "The Origins of Modernity: Caste and Feudality (India, Europe and Japan)" (1988).

154. See, for instance, White, *Machina Ex Deo* (1968); Mann, *The Sources of Social Power* (1986), "European Development" (1988); Hall, *Powers and Liberties* (1985); Baechler, "The Origins of Modernity" (1988); K. F. Werner, "Political and Social Structures of the West" (1988); and Hallam, "The Medieval Social Picture" (1975).

155. Hall, *Powers and Liberties* (1985), p. 135.

156. Mann, "European Development" (1988), p. 12.

157. Hallam, "The Medieval Social Picture" (1975), p. 49.

158. K. F. Werner. "Political and Social Structures of the West" (1988), p. 172. This German medievalist argues that the central force underlying the "miracle" was Christianity. He refers here to the Catholic church, to the body of doctrine (Catholic and later also Protestant), to the social and political institutions that were influenced both by church and by doctrine, and to the faith of European people, which, in Werner's opinion, had much to do with their innovativeness, their sense of "restlessness," their rationality. Werner acknowledges that many causal factors were at work in bringing about the rise of Europe, and he does not hesitate to claim a role for the natural environment. But it is clear that his theory is primarily based on

religion. And beyond this, one senses that this scholar himself may see the hand of a Christian god in the rise of Christian Europe.

Werner first of all makes a strong case to the effect that European history maintained a continuity of institutions, and of progress, from the time of the Roman Empire down through the Middle Ages, and that the late Empire established both ecclesiastical and lay institutions which continued down through the Middle Ages and gave that era its character. Chief among the institutions is the Catholic church. Werner wishes to depict the church as having a determining influence on history from the time of its founding. He sees the rise of Europe as a process that was guided throughout by the Christian religion, as institution and doctrine. If Werner were simply presenting a theory of history that accords religion a dominant role, I would not be discussing his views in this book. I would agree that its role has been underemphasized by historians, conservative no less than Marxist. The reason I deal with Werner's views is that he makes clear his belief that it is not religion in general but the Christian religion which played the historically efficacious role in the European "miracle." Werner's views are distinctly Eurocentric. Perhaps the key comment is the following:

> [If] we had to choose one word that, by itself alone, were capable of expressing an essential factor in what we understand by the "European miracle," we would choose the philosophical term . . . "unrest" . . . "restlessness" . . . *Ruhelosigkeit* . . . Whilst Asia and its wisdom and, in its train, the great religions and philosophies, strive toward the art of . . . seeking out the centre of the soul and of the world, and of resting in God, of having *arrived*, the European of the European miracle is a man who is always ready to take off once more, once he has arrived . . . But where must be sought the *causa causans* of this mentality? The spur to anxiety is to be seen in the pangs of sin . . . in the search for pardon and grace. The importance of the deliverance of the soul gave an hitherto unheard-of prominence to the individual, independent of his social rank, the individual who . . . [does not abandon himself] to destiny . . . [The] sense of responsibility seems to me to be one of the strengths of the Europeans that are to accomplish the "miracle" (p. 185).

Thus: Europe's religion, Christianity, instills a kind of mentality in "European man" which explains the basic source of the "European miracle."

The objection to this is the same one we put forward to all of the notions about Europe's supposedly unique "rationality." Whether that putative trait comes from religion (Werner) or from the post-Neolithic European tribes (Mann) or from any other source, the basic objection is: how can you *really* justify a statement that makes Europeans brighter, better, bolder than non-Europeans, if you accept the fundamental axiom that all human communities have the same ration of mind as all others? It is one thing, and indeed an unexceptionable thing, to credit the human mind with prime causality in culture change, through the innovation of social, technical, and purely intellectual things. But it is something quite different, and quite suspect, to credit the minds of human beings from *some* communities—and *not others*—with all of these qualities of innovativeness, restlessness, sense of responsibility, intellectual eagerness, respect for others, and so on, qualities usually summed up in the word "rationality." Europeans are rational but so too are non-Europeans.

159. See, for example, Palmer, *Atlas of Modern History* (1957); Bjørklund, Hølmhoe, Røhr, and Lie, *Historical Atlas of the World* (1970); and Kinder and Hilgemann, *The Anchor Atlas of World History*, vol. 1 (1974).

160. I discuss this in *The National Question* (1987b).

161. Padgug, "The Problem of the Theory of Slavery and Slave Society" (1976).

162. Baechler, "The Origins of Modernity" (1988).

163. Baechler, in "The Origins of Modernity" (1988), p. 39, suggests that we should "juxtapose the barbarous Europe of the Halstatt period [about 600 B.C.] . . . and Africa on the verge of colonization in the nineteenth century [A.D.]."

164. Baechler finds a true aristocracy also in Japan but believes that Japan failed to emulate Europe for various other reasons.

165. Baechler considers it quite unremarkable that the political chaos of the Dark Ages gave way smoothly to strong states in Europe. "Inevitably" there will be the "reconstitution of larger polities" (Baechler, "The Origins of Modernity" 1988, p. 50). But it is equally inevitable *for India* that the political chaos of 1000 years ago will *not* be cured and permanently, thereafter, the "polity . . . is not a reality in India" (p. 45).

166. Baechler, "The Origins of Modernity" (1988), p. 59.

167. Baechler, "The Origins of Modernity" (1988), p. 53.

168. Baechler, "The Origins of Modernity" (1988), p. 56.

169. Baechler, "The Origins of Modernity" (1988), p. 45.

170. Baechler, "The Origins of Modernity" (1988), p. 53.

171. Godelier, *Sobre el Modo de Producción Asiático* (1969), p. 58.

172. Brenner, "The Origins of Capitalist Development: A Critique of Neo-Smithian Marxism" (1977); "Agrarian Class Structure and Economic Development in Pre-Industrial Europe" (1985, originally published 1976), and "The Agrarian Roots of European Capitalism" (1985). After the 1976 article first appeared in the journal, *Past and Present*, a series of critiques was published in that journal, and Brenner replied in 1982 with the article, "The Agrarian Roots of European Capitalism." A volume, *The Brenner Debate: Agrarian Class Structure and Economic Development in Pre-Industrial Europe*, containing the two *Past and Present* articles and several critiques, and edited by Aston and Philpin, appeared in 1985.

173. In my opinion the popularity of this thin theory is due principally to two things. First, put forward as a Marxist view, grounded in class struggle, it proves to be, on inspection, a theory that is fairly conventional, if somewhat rural in bias. It seems to follows that class-struggle theories lead to conventional conclusions. And secondly, Brenner uses his theory ("The Origins of Capitalist Development: A Critique of Neo-Smithian Marxism," 1977, pp. 77–92) to attack the unpopular "Third-Worldist" perspectives of dependency theory, underdevelopment theory, and in particular three other neo-Marxists—Sweezy, Frank, and Wallerstein—who argue that European colonialism had much to do with the later rise of capitalism. Brenner is a thoroughgoing Eurocentric tunnel historian: non-Europe had no important role in social evolution at any historical period. Unaware that colonialism involves capitalist relations of production—see Chapter 4 below—he claims that the extra-European world merely had commercial effects on Europe, whereas the rise of capitalism was in no way a product of commerce: it took place in the countryside of England and reflected class struggle, not trade. See critiques of Brenner collected in Aston and Philpin (1985) by Hilton, Croot and Parker, Wunder, Leroy, Ladurie, Bois, Cooper, and others. Also see Torras, "Class struggle in Catalonia" (1980) and Hoyle, "Tenure and the land market in early modern England: Or a late contribution to the Brenner debate." (1990).

174. Taeuber, in Freedman, *Family and Kinship in Chinese Society* (1970).

175. (1) If only one family member is a wageworker, loss of employment is a disaster. If several are wageworkers, normally some will be earning an income when

others are laid off. (2) If we assume an ability to save a certain percentage of income, multiple income earners will provide a larger absolute amount of savings, that is, capital; and the absolute amount of capital can be critical in entrepreneurship. (3) Having kin to borrow from is useful for small-scale entrepreneurship. (4) Kinfolk can supply unpaid labor. These principles are well known in Third World communities.

176. Hajnal, "European Marriage Patterns in Perspective" (1965), pp. 101–146. This paper is one of the most widely cited publications on the subject of demography in the European miracle literature.

177. See note 37 above.

178. Hajnal, "European Marriage Patterns in Perspective" (1965), p. 101.

179. Hajnal, "European Marriage Patterns in Perspective" (1965), p. 134.

180. Hajnal concedes that he has only contemporary data for non-Europe, but merely suggests that historical data would prove his point even more clearly because modern non-European family patterns, in his view, are changing in the direction of European patterns: are being Europeanized. And when "all the qualifications about the data have been made, there can be no doubt that our original generalization remains" ("European Marriage Patterns in Perspective," 1965, p. 106).

181. Stone, *The Family, Sex and Marriage in England 1500–1800* (1977), p. 509.

182. Mann, *The Sources of Social Power* (1986), p. 408; Crone, *Pre-Industrial Societies* (1989), p. 152; Jones, *The European Miracle* (1981), pp. 15–16; Macfarlane, *Marriage and Love in England: Modes of Reproduction 1300–1840* (1986).

183. Laslett, "The European Family and Early Industrialization" (1989).

184. Taeuber, "The Families of Chinese Farmers" (1970), pp. 63–86.

185. See Freedman, *Chinese Lineage and Society* (1966), p. 49.

186. See, for example, Handler "Review of Macfarlane, A., *Marriage and love in England*" (1989); Hilton, "Individualism and the English Peasantry" (1980); Kertzer, "The Joint Family Household Revisited: Demographic Constraints and Household Complexity in the European Past" (1989); and Berkner, "The Use and Misuse of Census Data for the Historical Analysis of Family Structures" (1975), and "The Stem Family and the Developmental Cycle of the Peasant Household" (1989).

187. For instance, G. Lee, "Comparative Perspectives" (1987), p. 65, points out that "[Many] scholars contend that the majority of families in any society are and always have been nuclear, regardless of the cultural elements favoring extended families."

188. Macfarlane, *The Origins of English Individualism* (1978), chap. 1 and "The Cradle of Capitalism" (1988), p. 344.

189. See, for example, Hilton, "Individualism and the English Peasantry" (1980) and Handler, "Review of Mcfarlane, A., *Marriage and Love in England*" (1989).

190. "[It] seems likely that [primitive peoples'] patterns of behavior in . . . respect [to fertility and mortality] bore a strong resemblance to those which can be observed in many animals," Wrigley, *Population and History* (1969), p. 37. Wrigley is writing about present-day hunting–gathering–fishing peoples. A "description of the relationship between animal social conventions and the regulation of animal population numbers" is "a convenient point of departure for the study of primitive man" (p. 37).

191. Crone, *Pre-Industrial Societies* (1989), p. 153; Hall, *Powers and Liberties: The Causes and Consequences of the Rise of the West* (1985), pp. 130–132; Jones, *The European Miracle* (1981), pp. 3, 13–15, 217–19, 226–227, 231, and elsewhere; Laslett, "The European Family and Early Industrialization" (1989), pp. 235–240; Macfarlane, "The Cradle of Capitalism" (1988), chap. 14; Mann, *The Sources of Social Power* (1986), p. 408.

192. Hall, *Powers and Liberties* (1985), pp. 130–131.

193. Laslett, "The European Family and Early Industrialization" (1989), p. 237.

194. See Croot and Parker, "Agrarian Class Structure and the Development of Capitalism: France and England Compared" (1985). This essay, and others in the Aston and Philpin book, *The Brenner Debate: Agrarian Class Structure and Economic Development in Pre-Industrial Europe* (1985), discuss in detail the low level of landownership in medieval Europe.

195. Stone, *The Family, Sex and Marriage in England, 1500–1800* (1977), pp. 53–54.

196. Stone, *The Family, Sex and Marriage in England, 1500–1800* (1977), p. 652.

CHAPTER 3

Before 1492

In this chapter and the following one I will argue three broad propositions.

1. Prior to 1492, the progress toward modernization and capitalism which was taking place in parts of Europe was also taking place in parts of Asia and Africa. The basic process was hemispheric in scale. It was a process of change out of a precapitalist, agrarian form of class-structured society and toward a primitive form of capitalism. There was nothing teleological about this process; it was not some sort of evolutionary striving toward a foreordained goal, capitalist society. Merely, I argue, whatever happened in Europe *also* happened in other parts of the Eastern Hemisphere. I will use the word "feudalism" to describe the class-structured agrarian societies of Africa and Asia as well as Europe (and will give my reasons for using this word in this way). The later, emerging formation I will call "protocapitalism." In 1492, it is likely that more than half of each continent in the Eastern Hemisphere was basically dominated by a feudal social formation. Protocapitalist centers were rising in various parts of all three continents, and were interconnected in a single web or network, stretching from western Europe to southern Africa to eastern Asia.

2. This hemisphere-wide system began to break apart shortly after 1492, because of the wealth and power acquired by Europeans in America. America was conquered by Europeans, not by Asians or Africans, because of Europe's location on the globe, not because of any European superiority in level or rate of development or "potential" for development.

3. The massive flow of wealth into Europe from colonial accumulation in America and later in Asia and Africa was the one basic force that explains the fact that Europe became transformed rapidly into a capitalist society, and the complementary fact that Asian and African protocapital-

ist centers began to decline first in relative and then in absolute importance. Development began in Europe and underdevelopment began elsewhere. Many processes internal to Europe were important causes of
change, of development, in that continent, but the one basic process,
which ignited and then continuously fueled the transformation, was the
wealth from colonialism.

The first proposition is the topic of the present chapter; the second and
third are the topics of Chapter 4.

I will not and cannot demonstrate the truth of these propositions. I
will simply present a substantial amount of evidence that supports them,
and will show that the propositions fit in with other known facts in a
coherent theory—a theory that, I suggest, makes sense. That is as far
toward "demonstration of truth" as I can go, given the evidence of which
I have knowledge and the amount of detailed argumentation that can be
squeezed into this chapter. Some parts of this argument (such as the
pre-1492 development of Africa) will theorize well beyond the available
evidence, because, in my view, the facts needed to confirm or disconfirm
these parts of the argument have not yet been obtained, have not yet been
sought with sufficient diligence by diffusionist scholarship. For the most
part, however, the theorizing will be grounded in strong empirical
evidence. In addition to this evidence, there is the weight of evidence
presented in the last chapter *against* various opposing theories, those that
deny the importance of non-Europe before and after 1492. This has given
us, so to speak, a level playing field for considering the issues to be
discussed below.

MEDIEVAL LANDSCAPES

Before 1492, the various civilizations of Asia, Africa, and Europe were
very different from one another in many ways, but they were very like one
another in other ways. I believe that the ways in which they were different
did not have significance for cultural evolution.[1] In Chapter 2 I outlined
various theories that claim that particular differences between Europe and
other civilizations *do* explain the unique rise of Europe, and I tried to show
that these theories are unconvincing. In the following discussion I will
deal with some parts of culture that clearly are crucial for cultural
evolution, and I will try to show that the patterns found in medieval
Europe were *not* significantly different from the patterns found in other
civilizations. I will argue that modes of production, class structures,
systems of spatial exchange, and urbanization were broadly similar across

many civilizations, were evolving in much the same way, and to some extent were parts of a common hemisphere-wide process.

During the century or so preceding 1492, most of humanity lived in class-stratified agricultural societies. The great majority of people in these societies were peasant farmers, producing their own subsistence and forced to deliver a significant share of their output (or labor, or cash income) to an elite, or ruling class, a class that usually held claim to the land and almost always held both formal and real power over the peasants. What I have described here is a mode of production, that is, a complex of traits including material resources such as land, material culture (tools and the like), labor employed in production and distribution, social rules governing access to material resources and distribution of the output, and some related traits. For medieval Europe this mode of production is called "feudal." It is part of a larger concept, "feudal society." One of the important features of European feudal society was the nature of states and political power. Another was the culture of the landlord class, with its titles, its chivalry, and the rest. A third was the importance, in some regions and epochs, of serfdom. But underlying (or at any rate accompanying) these features was the general fact of feudalism as a mode of production, a landlord–peasant, class-stratified, agricultural society in which the landlord class was fed by surplus extracted (always with some degree of force) from peasant producers. This mode of production, with variations, was also a basic feature of almost all of the other class-stratified agricultural societies of the Eastern Hemisphere.[2] I will therefore use the term "feudal mode of production" for all such societies.

Others have used the term in this way but have encountered various important objections. Those scholars who insist that the peculiarly European features are, indeed, the evolutionary engines of change will naturally reject the description of other sorts of society as "feudal." Max Weber, for instance, thought that European feudal estates were unique and were crucial causes (or conditions) of progress. Those Marxists who consider serfdom to be a crucial feature in evolutionary terms would not want to use the term "feudalism" for societies that did not have serfdom (although many outside of Europe did).[3] Samir Amin rejects this broad usage of the term "feudalism," on the grounds that it tends to require of us that we use European feudalism as a model against which to measure other similar societies in other continents. Therefore, he prefers the term "tributary" to the term "feudal," arguing, correctly, that the various forms of surplus extraction in this mode of production (tax and rent; cash, labor, and product) can be assimilated to the concept of tribute paying.[4] My view is that Eurocentric historians do not have a copyright on the term "feudalism" and so it is not only valid but also in a sense just to use this term for the mode of production wherever we observe it, in any continent and any social formation. There remain other objections. What of the small urbanized societies found here and there across the map during this period? We will come to this matter in a later section of the chapter. How should we describe societies that are very aberrant from the basic landlord–peasant model? What about class-stratified pastoral societies? What about the class-stratified societies in which there is kinship linkage between producing class and ruling class? These matters of definition are important and I will try to deal with them in the context of the discussion.

There are many unanswered questions about the origins and evolution both of agriculture and of the feudal mode of production. Until recently most scholars believed that agriculture, class stratification, and many other attributes of civilization had originated in the ancient Near and Middle East. (We discussed this in Chapter 1.) Given this set of propositions, combined with explicit and implicit beliefs about the cultural backwardness and unprogressiveness of Asians and Africans, it was almost axiomatic that the agricultural landscapes of feudal Europe must have attained a qualitatively higher level of development—or, alternatively, must have had greater potential for rapid change—than those of many parts, and perhaps all parts, of medieval Asia and Africa. It seemed logical to believe that agriculture as such was still in the process of diffusing outward in some peripheral parts of the hemisphere during that period. For instance, as we noted in Chapter 2, historians tended to believe that most of southern Africa was "preagricultural" even in early modern times. Scholars speculated, and legitimately so given the basic model, as to the dates when agriculture in general, and each form of domesticated plant and animal in particular, had first reached each peripheral region in the general diffusion process.

This model began to crumble fairly recently. Very early dates for the Agricultural (Neolithic) Revolution began to appear for parts of Southeast Asia, dates of perhaps 9,000 years ago (the generally accepted idea is that agriculture in the Middle East is 10–12,000 years old). Pottery seemed to be just about that old in northeastern Asia and Japan. Soon afterward, very early dates for agriculture emerged for India, New Guinea, and other regions.[5] Today, although the majority view still seems to be that agriculture first arose in the Middle East, very many scholars believe otherwise. Many argue for independent and perhaps simultaneous origins in the Middle East and Southeast Asia; some would add West Africa. But it is also possible that the Agricultural Revolution occurred everywhere more or less at once.[6] By this I mean that the complex of crops, animals, tools, and ideas was being developed in many societies simultaneously (probably over a very long period), and each new trait tended to diffuse rapidly to those other parts of the hemisphere in which such a trait was a useful innovation, in an overall process that I call "criss-cross diffusion." This process gradually built up an agricultural landscape over a vast region of the hemisphere, extending (with unimportant gaps) across the entire swathe of tropical and midlatitude lands possessing moderately favorably climate and soil.[7]

In any event, it is now generally accepted that the diffusion of agriculture took place fairly long ago and by the Middle Ages agriculture had reached most of those regions in the Eastern Hemisphere in which environmental conditions are favorable for farming. Agriculture was still spreading at this time, but it could no longer be considered the Agricultural Revolution. Farming had been pushed poleward to a point not far short of its present latitudinal limits. In the Western Hemisphere the northern limit of maize in 1492 was not far south of the present limit of grain cultivation in central Canada. In both hemispheres almost all of the crops and livestock types that are important today had already been domesticated, although varietal improvement was still going on. As a generalization, it can be argued that each agricultural region had by this time selected for itself, from the long list of hemispheric domesticates, the combination of crops and stock best suited to its environmental conditions and cultures; most groups of related domesticated crop forms in any one part of the hemisphere were also known in many other parts.

One very dramatic bit of evidence in this matter was the swift spread of Western Hemisphere crops through the Eastern Hemisphere after 1492. This extremely rapid diffusion of maize, cassava, tobacco, sweet potato, white potato, and other crops, and the rapid way in which these domesticates became culturally important, shows how rapidly the diffusion of domesticates would occur when the process was one of the diffusion of previously unknown innovations: we can assume that most Eastern Hemisphere domesticates were no longer diffusing very rapidly. Where agriculture was spreading, it was taking place mainly in peripheral zones such as highlands, some forested regions, and remote islands, as a result mainly of social processes like migration, conquest, and land shortage.[8] In the years before 1492, agriculture was practiced, from southern Africa to northern Europe, northern Asia, southeastern Asia, and most regions of the Pacific Ocean, including Hawaii. For the most part, cultures we would describe as "nonagricultural" had chosen not to practice farming; they were not, therefore, "preagricultural."[9]

Probably the same holds true for the more complex forms of agricultural technology. The knowledge of irrigation, the plow, the use of fertilizers, complex rotations (including fallowless rotations), and other features of intensive agriculture had probably diffused by this time to all those parts of the agricultural landscape where farmers found it desirable to use them, either to increase output, to reduce labor requirements, to meet the demand for surplus delivery, or for any cultural reason whatever.[10] I would take the argument even a step farther. Throughout most of this landscape the diffusion of significant innovations had gone so far that the productivity of human labor was hardly ever limited by lack of technical knowledge of a kind available to other farmers in some other part of the hemisphere.[11] But this is speculation.

Agricultural societies are not always class stratified. But there is abundant evidence that most agricultural regions across the hemisphere displayed, in this period, a combination of agriculture and the landlord–peasant system of stratification, thus a mode of production I label "feudal." This point will be contested on two grounds. One of the objections, commonly heard from (some) Marxists, argues that medieval non-European agricultural modes of production were somehow lacking in the potential for change that we associate with the European feudal mode. This argument (the "Asiatic mode of production," "Oriental despotism," etc.) was discussed sufficiently in Chapter 2.

The second difficulty is a matter of the spatial pattern. Where, on the map of the medieval Eastern Hemisphere, do we find class-stratified agricultural societies, and where do we find classless agricultural societies? The answer must be given in two parts. First, we know beyond dispute that the class-stratified mode was dominant in nearly all agricultural regions of Asia, with clear patterns of landlord–peasant conflict. Arguments tend to focus on Africa. But there is little doubt that the landlord–peasant exploitative relation was dominant in much of northeastern Africa (for example, Ethiopia), the Sudanic zone from the Atlantic east beyond Lake Chad, some parts of the Lake Region of East Africa, southeastern Africa around the Zimbabwe imperial zone, and part of coastal East Africa. It is now known, also, that many of the forest-zone and dry-forest-zone states of West and Central Africa (Akan, Yoruba, Congo, and so forth), displayed this mode of production or something very like it, and research on the historical geography of this large region has just, in essence, begun.[12] Therefore the map of the feudal mode of production in Africa is very extensive. Second, I would argue (following Samir Amin) that nearly all state-organized societies were class societies, that the medieval state functioned in a tight relationship to the exploitative process and ruling-class politics. More than half of medieval Africa, in terms of area and population, was state-organized and therefore, I reason, more or less class stratified. I conclude, from this very sketchy examination of the medieval spatial patterns of agriculture, technically complex agriculture, and class, that the feudal mode of production dominated more than half of Africa, Europe, and Asia, and some parts of Oceania, in this period.

The ruling class in feudal societies is, almost everywhere, a landlord class, although the control of land by this class may take any of several legal forms. Some members of this class are bedecked with titles, but the distinction between nobility and gentry is not crucial in evolutionary terms and both forms (as well as others) were widespread across the hemisphere.[13] This class is, after all, self-perpetuating, and it may use inherited titles as a signal of class membership or it may use other devices to the same effect, or both. Indeed, membership in the nontitled gentry may, as in China at various times, improve a family's chances of retaining ruling-class status and wealth amid the changing winds of state politics. Nor is the distinction critical, in this context, between higher and lower grades of nobility, and between landlords and government officials (who likewise derive their wealth from land). As we discussed in Chapters 1 and 2, there is no substance to the traditional view that the European medieval landlord class somehow was closer to pure private land ownership than were the landlord classes of other places. Marx was wrong in accepting this traditional view, because he knew little about non-European class structures. Weber, likewise, was wrong in drawing a sharp distinction between the supposedly European pattern of seigneurial tenure, with land held firmly by the landlord under some sort of arrangement with higher-order lords and kings, and the "service tenures" which he thought to be characteristic of most other societies.

The distinction between hereditary and service tenures is very fuzzy. In Europe, service tenure was the typical form in strict terms (with grants conditional upon pledges of fealty, military support, etc.), but grants tended to become hereditary. Broadly, the same held true in other societies. Holders of fiefs or grants on service tenure might move from fief to fief (or hold a changing portfolio of fiefs), but the important point is that class membership permitted one to *hold* a fief, and to draw one's wealth from it (and its occupants), so long as one retained membership in the ruling class. In a crucial sense, property is private so long as an individual or kin group continues to hold valid control, and this was the case in many regions, in spite of periodic upheavals and replacements. But land can be called private in another sense, that of its value in a land market. But this implies a basically (or nearly) capitalist situation, found only in a few highly commercialized rural regions, European and non-European, before 1492.[14] The Chinese gentry, the Hindu fief-holders, even the Mughal *jagirdars* who had been granted fiefs on service tenure and quickly farmed them out, or converted them into private, heritable property, all displayed the classic features of a feudal landlord class.[15] The European feudal-era landlord class was not more advanced, more ready (as it were) for capitalism and modernity, than the landlord classes in many other regions.

The so-called European manorial system is sometimes said to have been a distinguishing feature of feudalism, a peculiarly European giant step toward private ownership and large-scale labor use, something largely absent from non-European areas and critical in the evolution toward capitalism. Large estates were widespread across the hemisphere, but the special organizational form of demesne farming by unpaid peasant labor was found in fewer areas. The manorial system in the narrow sense of the term, including coordinated demesne farming with corvée labor in gangs as well as peasant holdings, and with some manufacture along with agricultural production on the manor, was found in several areas outside of Europe. It was important in China and in southern India.[16] But demesne farming was not dominant throughout Europe (it was uncommon in the Mediterranean zone), bore no resemblance to capitalist agriculture, and in any case had nearly died out in western Europe by the fourteenth century. Hence the relatively stronger development of this trait in Europe than most other regions (such as northern India) cannot account for the transition, much later, to capitalism in one area and not the others.

Related to this question is the old European misdefinition of Indian villages, unfortunately accepted by Marx, as closed, corporate entities (hence, for Marx, as survivals of primitive communal society). The medieval Indian village did indeed have corporate characteristics; it did have communal control of usufruct (though not, apparently, communal ownership); and it did display the tight combination of farming and handicrafts which Marx found to be highly significant and seemed, to him, to explain the cohesiveness of the village, its ability to remain unchanged in the face of external shocks from colonial capitalism, yet, by the same token, to resist social transformation. But European villages also retained certain corporate characteristics, perhaps even more pronounced than those of Hindu villages, where caste communities correlated very poorly with village settlement patterns.[17] In this matter we may be confronting the classic error of telescoping history, perceiving the breakup of the European village *after* the rise of capitalism, and assuming therefore that these villages had been dissolving as corporate entities many centuries earlier. Furthermore, communal land ownership was relatively unimportant in this period for both India and Europe; the villages normally held only delegated rights (including the right to common land), which could be and sometimes were violated by landlords. The true owners of most of the productive land, holders of hereditary and transferrable estates, were, in both areas, the ruling class. Finally, the combination of agriculture and handicrafts was also present in European villages.[18] Apparently it dissolved well after 1492 with the rise of capitalism. In sum, although Indian feudalism was in no sense identical to the European variety (or varieties), it bore the same general characteristics as a mode of production and the same potential for evolution toward capitalism. This argument can probably be made also for many regions of Asia and Africa. The medieval European village seems not to have been very unusual among the array of village settlement and social forms across the hemisphere.

The producing class in feudalism consists, usually, of peasants, who farm the landlord's estate in household-scale units and provide labor, produce, or cash as rent. Serfdom is often thought to be the characteristic labor form of feudalism, on the European model.[19] Serfs of the basically European sort were indeed found here and there in Africa and Asia, although the specific history of enserfment in late-Roman Europe was unique and its legal form was rarely encountered elsewhere. What we find, rather, is a panorama of forms of unfree labor, that is, labor of peasants tied somehow to the landlord's estate, through all three continents.[20] On the other hand, some scholars (among them Brenner, a Marxist, and Baechler, a conservative) rather idealize the European peasant of the fourteenth and fifteenth centuries in western Europe and see in that person a freehold farmer, imbued with the entrepreneurial spirit and so forth.[21] This is again a telescoping of history. Those peasants were tenants, still tied to estates in manifold ways; not until later times, well after 1492, was there a strong emergence of an important freehold, capital-accumulating, kulak-style class, ready for rural capitalism. The European peasant was not particularly unusual. Peasants who were forced to give labor service, or product, or cash, as rent or tribute or tax (paid to the landlord), who were not free to move from the landlord's domain, and whose status was inherited by later generations, were found in many parts of medieval Asia and Africa as well as Europe.

There was a measure of interconnectedness among the feudal agricultural societies, enough to suggest that we should think of the whole hemisphere-wide zone of class-based agricultural societies not as separate social entities but as a single feudal landscape with regional variations that sometimes included sharp boundaries and sometimes did not. Clearly there was a great deal of criss-cross diffusion among these regions, as evidence, for instance, the commonality of agricultural techniques over large areas. (The claim made by some European historians,[22] that medieval European agriculture was unique in technological level and thus somehow ignited progress toward capital-ism, is invalid, as we discussed in Chapter 2. European agriculture shared most traits with other regions and was not uniquely advanced or peculiarly pregnant with social change.) It seems likely that the evolution of feudalism over much of the hemispheric landscape involved a steady deepening of the oppression on peasants, as more and more surplus was demanded, and the response by peasants included technological development and borrowing (diffusion), as well as migration toward peripheral regions and toward towns. At the same time, the ruling classes, as they exhausted the potential of their own subjects to increase surplus delivery, tried to conquer and exploit other communities of producers, to acquire external as well as internal fields of exploitation, and this also led to further interconnectedness of regions.[23] Yet, at the same time, feudal ruling class communities were united in webs of kinship, or bureaucracy, or caste, which sometimes extended over very large areas. We know that the neat parceling of societies into nation–states did not exist in those times, that language regions were ill-defined and language barriers of little significance, even that religious differences did not set up barriers to the movement of ideas, things, and people. Thus we should think of all (or most) feudal societies as sharing a common space, through which social forces and pressures diffused in all directions, over great distances, easily crossing the boundaries of states. Given this conception, it is not difficult to understand why the general evolution of feudalism as a mode of production was proceeding in about the same way over much of the hemisphere.

In the late Middle Ages there were signs of profound change in many agricultural regions of all three continents. There were indications of two sorts: signs of decay, or even imminent collapse, in the feudal system, and signs of change toward commercialized agriculture and toward rural capitalism. Throughout much of the hemisphere, the mode of production appears to have been in a state of decay, and we find increasing exactions, peasant revolts, migrations to agricultural frontiers and towns, intense warfare among ruling classes for access to producer populations, and more. By the fourteenth century, feudalism had entered a stage of crisis—although not of collapse—in Europe, but it appears that there were similar crises in parts of Asia and probably—as we will doubtless learn from further research—Africa.[24] In all three continents there was a movement of peasants to the towns, perhaps at roughly comparable rates. In no large region, European or non-European, could this have become a flood of rural–urban migrants, since urban population was still a small percentage of total population everywhere at the end of the fifteenth century. Still, it was an effect of crises in the rural areas. Whether these crises were indications that the mode of production was truly near collapse, and this from internal contradictions, perhaps cannot as yet be decided, but in any case feudalism in Europe was no closer to its final demise, prior to 1492, than were the feudalisms of many extra-European regions.

At this point in the argument, a disclaimer and a speculation. I am not arguing that the landlord–peasant mode of production had somehow gone through its allotted historical span and was about to collapse, or to transform itself into capitalism. The question whether capitalism had its earliest growth in an urban setting or a rural setting is a very complex one indeed; I will discuss the matter further below but I do not propose any sort of general historical theory of causation. I will argue only that the transition, or decay, or whatever one wants to call it, was far from complete in 1492, and the wealth from America precipitated the rise of capitalism and, simultaneously, the final decay of the feudal mode of production in Europe.

I speculate as follows: Given our overall model of an extremely rapid criss-cross diffusion of the cultural traits of agriculture (crops, stock, tools, water-management systems, etc.), and given the parallel conception of tight and intricate interlacing among class-organized agricultural societies in the medieval Eastern Hemisphere, one would expect that the general growth and evolution of the class-stratified agricultural form of society would proceed in a relatively *even* manner from one region to the next, as traits diffused, as social pressures were transmitted in space by migration, conquest, and the like, as ruling-class alliances proliferated, and so on. Perhaps the evolution of this feudal mode of production was everywhere conditioned by one common social fact: the steady and unrelenting demand of the landlord class and its allies (merchants, nobility, etc.) for more and more wealth, a demand that translated into constant pressure on peasants to increase production so that they could increase delivery of surplus. I view this as a long-term secular trend that led to specific responses in the peasant sector, including technological development, criss-cross diffusion of technology, assarting and pioneering, peasant revolts, rural–urban migration, participation in ruling-class military adventures, and more. I speculate, then, that these mechanisms evened out the social tensions that were created in many places by the increasing ruling-class demand for delivery of surplus. This would allow us to argue that, if the mode of production was in decay or in crisis in one part of the hemisphere, very likely the same was the case in many other parts of the hemisphere. In a word: the mode of production rose and then ebbed on a hemispheric scale, and what was happening in Europe in 1492 was also happening in Africa and Asia.

But why would we expect the feudal mode of production to decay or decline? This is the final point of speculation. We cannot assert simply that feudalism is a "stage" of evolution, and must eventually give way to the next, higher "stage" of evolution (capitalism), as some mechanistic Marxists argue. Nor can we accept the conservative form of this argument, which sees feudalism giving way to a higher and more "modern" form of society (capitalism) as a result of humanity's inevitable forward progress, social, intellectual, and moral. Nor can we invoke a Malthusian force of inevitably heightening population pressure (a thesis which was shown to be false in Chapter 2, on grounds mainly that human cultures always control their demographic behavior, more or less rationally). I would propose the following explanatory model. There are two essential facts about this form of society: first, the fact of family-scale farming as a way of life; second, the fact of a landlord class extracting, or trying to extract, an ever-increasing absolute surplus from farmers. Peasant farmers respond to this pressure in many ways, as we noted above. They certainly try to increase their population so long as each additional human being in the community can produce the requisite surplus, that is, contribute labor which yields more production than is needed for consumption by the incremental member of the community and for that individual's contribution to surplus delivery. Certainly they try to add additional land for cultivation, and sometimes try to move to another location, seeking agricultural or other land. But mainly they *intensify*. That is, they increase agricultural productivity by continuously experimenting with new crop varieties, new tools, new techniques, and they are ever on the alert for news about innovations that have been tried successfully elsewhere—in the next village, the next valley, the next island.

The process of technological improvement has no limit, but a point will be reached when the *rate* of increase in labor productivity declines, generation by generation, century by century. Doubtless the rate was at its highest during the period when many new crops and stock types were being domesticated at a rapid rate, and when the main tools, and iron, were being brought into the system. By the Middle Ages, the rate of improvement overall would have declined to a level insufficient to permit farmers to meet the landlords' incremental demands for surplus. If we set aside some of the alternative responses, such as pioneering and rural–urban migration, which must have been available in some regions but not in others, we are left with the following situation: a general crisis in the feudal mode of production.

Now this discussion has been grounded in one assumption: that improvements in agricultural production are taking place through innovations mainly on the farm itself. This is largely the case for family-scale farming in medieval and premedieval times. Of course, production is also improved by importing water and nutrients into the farm through irrigation or drainage. And always there is some off-the-farm sale or other exchange of products, and sale or exchange of products from the farm for inputs like fertilizer, seed, and labor. So the individual peasant farm, or family farm, is a relatively but not absolutely self-contained microgeographic system. We know very well, and farmers in those days also knew very well, that the best strategy for engineering a dramatic increase in production from a microgeographic system like the peasant farm is to integrate it more fully into a larger, macrogeographic system. Mainly this involves increasing the input of water and fertility elements, like lime and manure, and changing the pattern of crops and stock from one which must primarily feed the farm family to one which can involve some specialization in products that are saleable and that are well-suited to the ecological conditions of the farm. (This would commonly mean some specialization in one or a few food products, which are both sold and consumed, or specialization in an industrial product, like cotton.) When, today, we speak of the "agricultural revolution" of recent centuries, we are describing a revolution at this macrogeographic level: modern family farms import huge amounts of fertility; they import (purchase) many tools, pesticides, and like elements which are produced elsewhere; they use substantial amounts of nonfamily labor; and they specialize in ways that (sometimes) involve ecological optimization. This list pretty much exhausts the revolutionary changes that occurred prior to the present century, and it suggests that internal, microgeographic improvements—which never ceased to take place—played a secondary role at this stage in the development of agriculture.

For these macrogeographic improvements to take place there must be a high level of *commercialization* of farming, because moving things into and out of the farm microsystem (or at any rate the village microsystem) is mainly a process of buying and selling. It would seem to follow that a general crisis of the feudal mode of production would have one of two possible outcomes. One of these is a relatively smooth transition to an economy in which there is massive off-the-farm cash demand for farm products and supply of purchasable inputs, along with cash payment of rent (or payment on shares to a landlord who then markets the share for cash). This scenario takes place in a landscape in which there is a large nonagricultural population, hence a landscape that is either urbanizing or participating in major long-distance trade. Stated differently: the crisis can be met if commercialization and urbanization are taking place. Alternatively, there can be revolutionary changes of another sort: peasant revolts, either mild (such as withholding of rent) or violent, major cultural transformation in social, political, or religious life, or something else equally revolutionary. Perhaps both alternatives must occur in some combination for the crisis to be resolved. I conclude that the rather clear pattern in which feudal contradictions intensify, as a result of increasing demands for surplus and decreasing ability of farm families to increase the level of surplus, must lead to a revolutionary change of some sort. Most of the changes that I have mentioned did, in fact, occur in Europe in the late Middle Ages and involved the overthrow of feudalism as a political and social system and its replacement by the modern system after the model of England's "Glorious Revolution." But I am *not* arguing that this rural set of processes *explains* the rise of capitalism. Certainly it *contributed* to the rise of capitalism, and specifically to the processes of increasing urbanization and increasing long-distance commodity movements which characterized the late Middle Ages throughout the hemisphere, processes which I label "protocapitalist," and which we discuss in the following section of this chapter.

PROTOCAPITALISM IN ASIA, AFRICA, AND EUROPE

I use the word "protocapitalism" not to introduce a technical term but to avoid the problem of defining another term, "capitalism." Obviously, the kind of economic system that we ordinarily think of as capitalist did not exist in the Middle Ages; we are dealing with its forebear, which (as I will argue) exhibited most of the basic traits of capitalism, but on a spatially and socially small scale, and generally within, or on the edge of, a much larger, dominant economic system associated with the feudal mode of production. Protocapitalism, therefore, is incipient capitalism, or near-capitalism, or adolescent capitalism. It is the system as it existed prior to the two revolutionary transformations which brought modern capitalism into existence. The first of these was the political transformation which is usually, and conventionally, called "the bourgeois revolution" or "bourgeois revolutions"—the creation of large polities that were dominated, not by the feudal landlord class, but by an elite of townsmen (burghers, bourgeoisie) and their entrepreneurial allies in the countryside. The most famous example, and in a way the defining case, was Britain's "Glorious Revolution" of 1688, and I will use the date 1688 as the symbol or token for the political triumph of capitalism. The second transformation was, of course, the Industrial Revolution, which did not really begin in a big way until the last quarter of the eighteenth century. In Chapter 4 we will examine the role played by colonialism and non-Europe in both of these transformations.

In all three continents we find relatively small rural regions (they were generally hinterlands of major port cities) along with a few highly commercialized agricultural and mining regions, which were clearly being penetrated by capitalism—were protocapitalist—in the period just prior to 1492. Among these were Flanders, southeastern England, northern Italy, sugar-planting regions of Morocco, the Nile valley, the Gold Coast, Kilwa, Sofala (and hypothetically part of Zimbabwe), Malabar, Coromandel, Bengal, northern Java, and south-coastal China. Land was owned by commerce-minded landlords or by urban protocapitalists.[25] Rents were generally paid in cash except in those areas, like Fukien, where more money profit could be extracted by landlords if they collected the farm produce and sold it themselves.[26] Agricultural production was organized in various ways, ranging from peasant-scale farming to plantations, and very significant quantities of a number of agricultural products were grown, sold, and exported: rice, cotton, sugar, pepper, etc. Industrial production was spreading out into the countryside in all three continents: the early putting-out system was actually de-urbanizing industry in northwestern Europe, as the control by guilds became loosened; probably the same was occurring in parts of Asia and Africa (where merchant and artisan guilds were also well developed and strong in the Middle Ages).[27] Over a much broader area, commodity production had fully penetrated the agricultural economy, and it is extremely doubtful whether west European peasant agriculture was more highly commercialized than that of many parts of China and India, as well as some other extra-European regions. Probably we can assume that level of urbanization is a good comparative indicator of level of agricultural commercialization for this period, since it must represent the main off-the-farm demand for agricultural products. By this measure, Chinese and Indian agriculture would have been more highly commercialized than European agriculture, because a larger percentage of total population was urban in those regions.

Cities dotted the landscape from northern Europe to southern Africa to eastern Asia. Some of these cities were seats of power for major feudal societies. Others were socially and geographically marginal to these societies, and were usually to be found along sea coasts, where they had mainly an interstitial relationship to the larger feudal societies, moving and trading goods among them and producing manufactured commodities for them. Probably it would be incorrect to speak of two distinct classes of urban place, internal and marginal (or peripheral), because many variations and gradations existed, and also because the internal, seat-of-power cities were in many cases also major centers for intersocietal trade and for nonagricultural production. Nevertheless, we can distinguish a special group of cities that were strongly oriented toward manufacturing and trade, were more or less marginal to powerful feudal states (some were within these states; some were small city-dominated states or even city–states), and were heavily engaged in long-distance maritime trade. Cities of this sort stretched around all of the coasts of western Europe, the Mediterranean, East Africa, and South, Southeast, and East Asia. In these cities the mode of production could probably be best described as incipient capitalism, protocapitalism—certainly it was not feudalism—with wageworkers being, apparently, the largest working-class sector, merchants, merchant–landlords, or merchant–manufacturers the ruling class, and economic activity a mixture of trade (movement of commodities, banking, and so on), manufacture (both large- and small-scale), and commercial agriculture.

Some of these mercantile–maritime cities were quite small, others quite large, but it appears that most of them were at roughly the same level in the development of protocapitalist institutions, classes, and technology. This is not surprising since they were intimately connected to one another in a tight network of trade, along which ideas, techniques, goods, and people flowed in all directions, in constant criss-cross diffusion.[28] (For example: Malacca, when the Portuguese first arrived, was trading with the Mediterranean, Inner Asia, East Africa, the Middle East, India, China, and probably Japan as well as all of Southeast Asia. The chronicler Tomé Pires assures us that, at the beginning of the sixteenth century, 84 different languages are spoken in that city, and, boosting its importance for the Portuguese, asserts that "whoever is lord of Malacca has his hand on the throat of Venice."[29] A second example from a much earlier period: the Tenasserim port of Kalah, in the tenth century, was trading with China and Arabia. According to Ibn al-Faqih, the parrots of Kalah talked in Persian, Arabic, Chinese, Indian, and Greek.[30])

The network of mercantile–maritime centers stretched, like a string of pearls, from the Baltic to the eastern Mediterranean, and from there southward to Sofala (or beyond—the history of East and southern Africa is still buried in colonial slumber) and eastward to Japan. The network also extended inland in all three continents, but the mercantile–maritime cities and oceanic routes were eventually of greater evolutionary importance in the rise of capitalism than were the inland centers. This was true for two (main) reasons.

First, foreign trade was the most peripheral of protocapitalist activities; it was literally beyond the reach of the law. (Inland cities that bordered on deserts would also have had this peripheral quality to some extent.) Thus, a protocapitalist port city could move products to and from any other oceanic port without having to pass through state-organized territories, and thereby avoid paying tolls, being forced to buy and sell goods to foreign merchants at intermediate trading centers, or perhaps even being denied permission to enter a state. It is worth noting, in this regard, that a substantial part of the high cost of Asian spices in European markets before 1492 resulted from the fact that shipments coming from India via overland routes ordinarily had to be passed from merchant to merchant at several trading points enroute, with profit taken at each intermediate market. The cheapness of Asian spices carried by the Portuguese in the sixteenth century therefore reflected, to a considerable extent, the fact that the spices could be on-loaded at an Asian port, and then transported direct to a European port with no intermediate transactions; perhaps this factor was more important than the generally lower cost of sea transport over land transport (a factor that is often overemphasized).

Second, long-distance commodity movement by sea, involving as it did the transport of vital staples as well as luxuries, was, among protocapitalist activities of the late Middle Ages, perhaps the closest we get to industrial capitalism in the urban economy of that time. It involved not merely an exchange of commodities but the production of many commodities including ships, and incorporated sophisticated technology, a large work force, complex transactions, and massive capital accumulation. This matter brings us back, inescapably, to the problem of defining "protocapitalism."

There is a widespread tendency, often encountered among Marxists but by no means confined to that school of thought, which argues the following position. Money, cash exchange, and trade have been going on for millennia but they do not signify capitalism or even the seeds of capitalism. This is so because capitalism is a matter of *production*, not *exchange*. "Real" capitalism requires the application of wage labor and the production of commodities. Exchange is merely buying and selling; it does not add value. For Marx, it produces wealth mainly as a result of unequal exchange (higher prices in one market than another, and the like), not as a result of labor input and the production of use-value.

From this model come a series of highly important theses. One is the argument that medieval European towns were not central to the rise of capitalism because their main activity was trade, exchange, not production. Therefore the rise of capitalism must have occurred, not in medieval towns but in medieval agriculture.[31] But a second thesis is more crucial for the issues discussed here. This is the argument that starts out by conceding that the great medieval trading cities and trading routes of Asia were much more impressive in scale than those of Europe and the Mediterranean, but this did not make them more significant for the rise of capitalism—because it was production, not exchange (trade, commerce), that was the crucial process. No matter how highly developed the trading routes and cities of Asia were, Europe's feudal agricultural production (in this argument) was closer to capitalism than either the rural or urban production systems of non-Europe and it is *this* fact—the nature of European rural society as contrasted with non-European rural society—that is crucial in explaining why capitalism arose in Europe, not in Asia (or Africa). The fallacy regarding rural production was discussed previously. But equally fallacious is the idea that Asian (and African) port cities, mercantile–maritime centers, were somehow purely or largely concerned with exchange, with "commerce." Here there are in fact three errors. First, production involves not merely change of form but also change of place. It is metaphysical to argue that there is something ontologically distinctive about the process of shaping nature into a "thing," a commodity. When a farmer produces an agricultural "thing," he or she must not only grow it but also transport it from field to farmstead and then to market, and must also transport inputs of water or fertilizer or labor from outside the farm. Farm production, therefore, involves both change of form and change of place. An automobile assembly line is a process of change both of form and of place. Thus, overall, *spatial movement is part of production*. It has nothing whatever to do with the entirely distinct process by which commodities are purchased and sold. Indeed, the farmer's crop can be subject to exchange right on the farm as well as in an off-the-farm market. Therefore, the medieval activities involved in moving commodities over long distances were not, ontologically, "exchange"; they were *spatial transport*. They involved huge labor forces, massive capital investment, major technologies—of navigation, ship construction, banking and insurance, and more—and significant tonnages. They produced use-value at the destination from commodities that had none, or less, at the point of departure. In a word, what is called "medieval trade" was a complex process in which production and manufacture played as great a role as did exchange.

The second error is the idea, very widely held today among historians, that the cities, the commodity movement, and the rest of the complex, was somehow a trivial process involving only the moving of a few luxury items to a tiny ruling class. In fact, most of the medieval seaborne trade was a matter of staple commodities, things like crude textiles, iron implements, rice, wheat, lumber, ships (which often were sailed from the place of construction to some other port where they were sold), and the like. But beyond that, the tonnage and value of products that would not be considered staples, things like pepper, sugar, finer textiles, pottery, and so forth, was, in and of itself, immensely important, because the market for such products was very large: the medieval elites were by no means insignificant.

The third error is a failure to perceive how important industrial production was in these medieval cities and their hinterlands. Thus, I conclude that the medieval mercantile–maritime system was very much a nursery bed of capitalism, in Asia and Africa as well as Europe.

The protocapitalist port cities of Europe were not more highly developed than those of Africa and Asia in the fifteenth century. This holds true regardless of the kinds of criteria chosen as measures. European cities, first, were not larger in absolute or relative population. In fact, urbanization in Europe was probably less advanced than urbanization in China, India, the Arab region, and no doubt many other non-European areas. The urban population in early Ming China was perhaps 10% of the total population.[32] In the Vijayanagar Empire of southern India it must have been at least as high: the inland capital alone held about 3% of the population—comparable centers in Europe, such as Paris, may have had half that percentage—and the coastal port cities were both numerous and large.[33] Second, the development of the techniques of business was fully as advanced, fully as complex, and fully as wideflung in space among the merchants and bankers of Asia and Africa as among those of Europe. (Tomé Pires said of Gujarati businessmen in 1515: "They are men who understand merchandise; they are . . . properly steeped in the sound and harmony of it" and "those of our people who want to be clerks and factors ought to go there and learn, because the business of trade is a science."[34]) Third, the technical and material means of production seem to have been at about the same level of development in many mercantile–maritime centers of all three continents, allowing for differences in the volume of production and trade, the kinds of merchandise, and the like. Maritime techniques were also comparable across the hemisphere: though they differed from ocean to ocean, it cannot be said that ships of one ocean were technologically more advanced than those of the others.[35] Manufactures in port cities and other industrial centers of Europe, Africa, and Asia were also roughly comparable in gross scale and level of development.[36] Fourth, the urban class composition of Asian and African centers appears to have been similar to that of European centers: in all regions there existed a powerful class of protocapitalists and a wage-earning class of workers, with or without involvement also of other classes such as feudal landlords, slaves, and so on. And finally, the old European myth, codified by Weber—that European cities were somehow more free than non-European cities, which were under the tight control of the surrounding polity—is essentially an inheritance from classical Eurocentric diffusionism, which imagined that everything important in early Europe was imbued with freedom while everything important in Asia (not to mention Africa) was ground under a stultifying "Oriental despotism" until the Europeans arrived there and brought freedom. The so-called "free cities" of central Europe were hardly the norm and were not central to the rise of capitalism. The partial autonomy of many mercantile–maritime port cities of Europe, from Italy to the Baltic, was of course a reality, and usually reflected either the dominance by the city of a relatively small polity (often a city–state) or the gradual accommodation of feudal states to their urban sectors, allowing the latter considerable autonomy for reasons of profit or power. But all of this held true also in various parts of Africa and Asia. Small city–states were common around the shores of the Indian Ocean, in the Maghreb, and in Southeast Asia; also common were quasi-independent cities, giving loose allegiance to larger states. This point was discussed in the previous chapter.

The preceding discussion was not a theory of the rise of capitalism. My aim was simply to show that all of the theories that claim causal superiority for Europe on the basis of Asia and Africa's supposed lack of progressive urbanization or because extra-European urban processes were not important since urban processes in general were of minor importance compared to rural processes—European rural feudalism—are very unconvincing.

It is not an exaggeration to describe this entire network of mercantile–maritime cities as a single protocapitalist system.[37] The surrounding space of class-organized agricultural societies was, as I argued previously, made up of separate societies and polities in separate regions but was, nonetheless, integrated enough so that persistent criss-cross diffusion and other movements led to a degree of unity; perhaps even a degree of intercontinental equilibrium. The unity was very much more intense for the network of protocapitalist cities. The image I have in mind for this is a network of strings of electric lights of various sizes and colors illuminating a garden party. The current, so to speak, which flowed among those port cities consisted of human beings (sailors, workers, merchants, etc.), material things (commodities, ships, fertile seeds and cuttings of crops, musical instruments, and much more), and ideas—technological ideas, innovative social, economic, and religious ideas, and so on.

All of this is well known in qualitative terms but not fully so in terms of its intensity, its spatial extent, and, most critically, its *unity*. The entire system can be viewed as a single entity, so tightly integrated that there must have been rapid, almost instantaneous, criss-cross diffusion *throughout* the system of essentially every material or immaterial culture trait that is relevant to the economic and technical and ecological progress of this form of society. I believe it is an error, built into our way of conceptualizing cultures and cultural differences, to believe that the very profound differences of culture among the various societies that comprised this system would, somehow, have been reflected in a lack of integration across cultural boundaries in matters concerning the technical–economic–ecological dimension of culture. (Recall our discussion above concerning the distinction between evolutionary and nonevolutionary or partly non-evolutionary aspects of culture, in the theoretical tradition of anthropologists like Steward.[38]) In those times, differences of language did not seem to interfere with the quest for profit among merchants and other participants in this system. (Recall the Greek-speaking parrots of Kalah, the 84 languages spoken in Malacca.) Nor were differences of religion any great impediment (as has been amply documented for Muslim–Christian–Jewish trade in the medieval Mediterranean[39]). Certainly there were limited social networks, membership in which was a matter of religion or nationality or even kinship. Abu-Lughod has shown that the pattern of connections and distinctions produced a set of eight overlapping social regions—she writes of "The Eight Circuits of the Thirteenth-Century World System"—although her data and argument are consistent with my present thesis that all regions were in fact subregions of one protocapitalist system.[40] State boundaries do not seem to have played an important inhibiting role in the flows across the system, except in certain fairly limited periods when either political conflict or specific imperial policies and practices did, indeed, disrupt trade through one or another political partition; the truly nationalistic forms of capitalist enterprise become important much later; in fact, after 1492.[41]

The network, or system, seems to have evolved over a period of several centuries, mainly from the tenth to the fifteenth. Without attributing cause, I would emphasize the fact that the period during which this system grew most rapidly in scale and intensity was the period during which the technology of oceangoing shipping increased explosively, in what may be thought of as a (or the) Spatial Revolution. In the Agricultural Revolution, we do not know whether the technical–ecological transformation was cause, or effect, or both, in relation to social transformation, although most scholars tend to treat agriculture as cause and social change as effect, rightly or wrongly. In the case of the medieval Spatial Revolution, it is most likely that the technological–ecological aspect was more a reflection of the economic and social processes associated with emergent protocapitalism and urban development than a cause of the latter. Nevertheless, the medieval Spatial Revolution was in one critical way a sequel to the Agricultural

Revolution: it intensified spatial flows much as the earlier revolution intensified *in situ* production. This is not to say that earlier boat technology and earlier long-distance sailing out of sight of land was insignificant: the question is one of intensity.

Perhaps, as a final speculation, we might think of the Spatial Revolution as part of a larger process which was responding to the maturation and decline of the feudal mode of production. Certainly it is true that increasing commodity demand by elites was a major stimulus, but it is also possible that the emerging crisis of feudalism—the decreasing rate at which an absolute increase in surplus could be extracted from peasant producers, and the resulting stresses and strains—had much to do with the rise of the intercontinental protocapitalist system. In any case, the dramatic long-distance voyages of discovery of the later Middle Ages, voyages by Chinese, Indians, Polynesians, Europeans, and others, should be conceptualized as moments in a genuine Spatial Revolution.

Much of this is speculation beyond the empirical data. But we have important data about the parallels of development from one urban system to another, and from one trading region to another. We also have dramatic cases of almost instantaneous diffusion: for instance, the appearance of the cannon in the Mediterranean region and in China almost simultaneously; perhaps in the same decade.[42] For the argument of this book, the one crucial generalization is the following: It is not surprising that the processes that I have called protocapitalist were going on across the Eastern Hemisphere in the later Middle Ages. Africa, Asia, and Europe were about equally close to—or distant from—capitalism and modernity in 1492. After 1492, the pace of development quickened for Europe and slowed for Africa and Asia, because of the wealth brought to Europe from America.

NOTES

1. Concerning my use of the term "cultural evolution," see Chapter 1, note 14.

2. The discussion in this section of the chapter mainly deals with the Eastern Hemisphere. The Western Hemisphere will be discussed separately in Chapter 4.

3. See Dobb, *Studies in the Development of Capitalism* (1947).

4. Amin, *Unequal Development* (1976) and "Modes of Production: History and Unequal Development" (1985).

5. Kabaker, "A Radiocarbon Chronology Relevant to the Origins of Agriculture" (1977); Megaw, *Hunters, Gatherers and First Farmers Beyond Europe* (1977); Vishnu-Mittre, "Origin and History of Agriculture in the Indian Subcontinent" (1978). See the review in Blaut, "Diffusionism: A Uniformitarian Critique." (1987a).

6. Blaut, "Diffusionism" (1987a).

7. The assumption here is that agriculture itself was evolving because it was useful, but there is the corollary assumption that humanity realized this not just in one favored place but in many places and among many peoples. This should not be surprising, given the fact that agriculture is useful nearly everywhere today. But it is definitely contradictory to the assumptions of Eurocentric diffusionism.

8. Examples of this process include the eastward movement of the medieval agricultural frontier in the forests of eastern Europe and the reclaiming of swampland in Iraq. Cropland was being expanded in these and other ways in many regions. In addition, it seems certain that some societies that had not previously practiced agriculture were being pushed into smaller or less favorable regions and therefore were turning to agriculture as a means of increasing food production to meet the land shortage.

9. See R. Lee, "Art, Science, or Politics? The Crisis in Hunter-Gatherer Studies" (1992).

10. Water-control systems in farming, including irrigation, drainage, broad-based terracing, raised- or drained-field construction, and natural-levee adaptations are probably as old as agriculture itself, because (1) all farmers everywhere know the problem of moisture control (adding moisture when there is deficit; removing moisture when there is a surplus and a danger of root drowning); (2) all of these procedures are initially *small-scale* actions taken on an individual farm (recall our discussion in Chapter 2 of the fallacies of the "hydraulic theory"); and (3) there is direct archaeological evidence of very ancient (9,000-year-old) drainage systems in New Guinea (Golson, "No room at the top: Agricultural intensification in the New Guinea Highlands," 1977), and drained- or raised-field systems in tropical America (Denevan, "Hydraulic Agriculture in the American Tropics: Forms, Measures, and Recent Research," 1982). Thus we can infer the primitivity of irrigation and these other water-management systems: probably they are as old as the Neolithic, along with drainage and raised-field systems. All this shows that intensive technology had *already diffused* insofar as it was going to do so, and nonadoption reflected something other than lack of information. As we noted in Chapter 2, irrigation *systems* diffuse as a social process, associated with class society. Concerning the myth that the plow was not used in Africa, see Hopkins, *An Economic History of West Africa* (1973) and Onimode, *Imperialism and Underdevelopment in Nigeria* (1982). Note that plows are used in tropical agriculture very sparingly—mainly for some operations in rice paddies.

11. In classless societies, I speculate that the bundle of choices concerning crops, tools, field systems, labor input, and the like, led to roughly a common level of output per person, unaffected by differences in environmental quality over a great range of environments. In some areas very extensive systems like shifting cultivation would be used; in others, intensive systems, like wet-rice cultivation, would be used. But the productivity of labor in terms of product per hour input would tend to be about the same in this model for both intensive and extensive systems. This would hold true if two assumptions are accepted: (1) that rapid and thorough diffusion had taken place, and (2) that population was controlled by farming peoples so as to optimize the situation concerning output and leisure time. *None* of this would be true in a class society, where the constraints on technology and labor use are influenced profoundly by the demands and power of the ruling class.

12. See, for example, Kea, *Settlements, Trade, and Polities in the Seventeenth-*

Century Gold Coast (1982); Isichei, *A History of Nigeria* (1983); Rodney, *A History of the Upper Guinea Coast 1545–1800* (1970); A. Smith, "The Early States of the Central Sudan" (1971); Usman, *The Transformation of Katsina (1400–1883)* (1981).

13. See Blaut, "Colonialism and the Rise of Capitalism" (1989). It is also true that in all these societies there were parallel high-status groups, clergy, bureaucrats, military people, and so on, but there seems not to have been any case of a large, clearly feudal society—I exclude a few cases of small urbanized power centers in dry, pastoral regions, and a few large cities—in which wealth and status was clearly divorced from land ownership and from the surplus extracted from peasants.

14. Apparently such private (saleable) ownership of agricultural land was found mainly near important urban areas, ports, mining areas, etc. See Rawski, *Agricultural Change and the Peasant Economy of South China* (1972); Das Gupta, *Malabar in Asian Trade: 1740—1800* (1967); Nicholas, "Town and Countryside: Social and Economic Tensions in 14th Century Flanders" (1967–1968); Kea, *Settlements, Trade, and Polities in the Seventeenth-Century Gold Coast* (1982); Rodney, *A History of the Upper Guinea Coast* (1970); Usman, *The Transformation of Katsina* (1981); Sherif, *Slaves, Spices and Ivory in Zanzibar* (1987).

15. On the importance of hereditary fiefs and landed property in Asia, see, for example, Elvin, *The Pattern of the Chinese Past* (1973); Sharma, *Indian Feudalism, c. 300–1200* (1965); Fei Hsiao-tung, *China's Gentry* (1953); Fu and Li, *The Sprouts of Capitalistic Factors Within China's Feudal Society* (1956); Rawski, *Agricultural Change and the Peasant Economy of South China* (1972); Tung, *An Outline History of China* (1979); Liceria, "Emergence of Brahmanas as Landed Intermediaries in Karnataka, c. A.D. 1000–1300 (1974); Mahalingam, *Economic Life in the Vijayanagar Empire* (1951); Hasan, "The Position of the Zamindars in the Mughal Empire (1969); Raychaudhuri, "The Agrarian System of Mughal India" (1965); Yadava, "Secular Land Grants of the Post-Gupta Period and Some Aspects of the Growth of the Feudal Complex in Northern India" 1966). For Africa south of the Sahara (for which there is as yet only fragmentary evidence), see, for example, A. Smith, "The Early States of the Central Sudan" (1971); Mabogunji, "The Land and Peoples of West Africa" (1971); Kea, *Settlements, Trade, and Polities in the Seventeenth-Century Gold Coast* (1982); Isichei, *A History of Nigeria* (1983); Onimode, *Imperialism and Underdevelopment in Nigeria* (1982); FRELIMO, *Historia de Mozambique* (1971); Rodney, *A History of the Upper Guinea Coast (1970); and Usman, *The Transformation of Katsina* (1981).

16. On the manorial system of China, see Elvin, *The Pattern of the Chinese Past* (1973). For India, see Gopal, "Quasi-Manorial Rights in Ancient India" (1963); Mahalingam, *Economic Life in the Vijayanagar Empire* (1951); Yadava, "Secular Land Grants of the Post-Gupta Period" (1966); Yadava, "Immobility and Subjugation of Indian Peasantry in Early Medieval Complex" (1974). Indian historians recognize important differences between the Indian and European forms of the manor, however. In early Indian feudalism manorial labor had some of the characteristics of serfs, some of wage laborers, and some of tenant farmers. Early Indian feudal estates also seem to have been less autarkic and insulated than the stereotypic European manor.

17. Marx's view is set forth in "The British rule in India" (1979). Irfan Habib, in part following Kosambi (both are Marxists), writes of "the *creation* of the traditional Indian village, closed and self-sufficient" (my emphasis) between 200 B.C. and 650 A.D., in a process involving "ruralization of crafts" and somewhat planned settlement of landless people in villages: Habib, in "The Social Distribution of Landed Property in Pre-British India" (1965).

18. On the unity of agriculture and handicraft industry in medieval European villages, see, for example, Sylvia Thrupp, "Medieval Industry 1000–1500" (1972).

19. Dobb, *Studies in the Development of Capitalism* (1947).

20. Serfdom was not characteristic of all parts of medieval Europe. On unfree labor in Asia and Africa, see, for example, Yadava, "Immobility and Subjugation of Indian Peasantry in Early Medieval Complex" (1974); Levitzion, "The Early States of the Western Sudan to 1500" (1972); Elvin, *The Pattern of the Chinese Past* (1973).

21. See Brenner and critics in Aston and Philpin, *The Brenner Debate: Agrarian Class Structure and Economic Development in Pre-Industrial Europe*(1988); Brenner, "The Origins of Capitalist Development: A Critique of Neo-Smithian Marxism." (1977); Baechler, "The Origins of Modernity: Caste and Feudality (India, Europe and Japan)." (1988). See comments on Brenner and Baechler in Chapter 2 above.

22. including Lynn White, Jr. in *Medieval Technology and Social Change* (1968); Michael Mann, "European Development: Approaching a Historical Explanation" (1988); Perry Anderson, *Lineages of the Absolute State* (1974).

23. Blaut, *The National Question* (1987b), chap. 7.

24. For India, see, for example, A. Chicherov, "On the Multiplicity of Socio-Economic Structures in India in the Seventeenth and Eighteenth Century" (1976); I. Habib, "Problems of Marxist Historical Analysis" (1969); S. Gopal, Nobility and the Mercantile Community in India" (1972); Radhakamal Mukherjee, *The Economic History of India, 1600–1800* (1967); Ramkrishna Mukherjee, *The Rise and Fall of the East India Company* (1958); Jha, *Studies in the Development of Capitalism in India* (1963); Nurul Hasan, "The Silver Currency Output of the Mughal Empire and Prices in India During the 16th and 17th Centuries" (1969); Yadava, "Immobility and Subjugation of Indian Peasantry in Early Medieval Complex" (1974). For West Africa, see Kea, *Settlements, Trade, and Polities in the Seventeenth-Century Gold Coast* (1982). For China, see Harrison, *The Communists and Chinese Peasant Rebellions* (1969); Parsons, *Peasant Rebellions in the Late Ming Dynasty* (1970); Fu and Li, *The Sprouts of Capitalistic Factors Within China's Feudal Society* (1956).

25. Appadorai, *Economic Conditions in Southern India* (1936); Elvin, *The Pattern of the Chinese Past* (1974); Nicholas, "Town and Countryside: Social and Economic Tensions in 14th Century Flanders" (1967–1968), pp. 458–485; Rawski, *Agricultural Change and the Peasant Economy of South China* (1972); T. Raychaudhuri, *Jan Company in Coromandel* (1962).

26. Rawski, *Agricultural Change and the Peasant Economy of South China* (1972).

27. See Appadorai, *Economic Conditions in Southern India* (1936); Gernet, *Daily Life in China on the Eve of the Mongol Invasion* (1962); Habib, "Problems of Marxist Historical Analysis" (1969); Mahalingam, *Economic Life in the Vijayanagar Empire* (1951); K. Nilikanta Sastri, *A History of South India* (1966); Tung, *An Outline History of China* (1979); Kea, *Settlements, Trade, and Polities in the Seventeenth-Century Gold Coast* (1982), Sherif, *Slaves, Spices and Ivory in Zanzibar* (1987).

28. Blaut, "Where Was Capitalism Born?" (1976).

29. Pires, *The Suma Oriental* (1944 edition).

30. Pires, *The Suma Oriental* (1944); Di Meglio, "Arab Trade with Indonesia and the Malay Peninsula from the 8th to the 16th Century" (1970). The location of Kalah is tentatively placed in the Mergui region: see Wheatley, *The Golden Khersonese* (1961).

31. This thesis is central to the famous debate over the role of urbanization in the medieval rise of capitalism; in Marxist literature this view is associated with Maurice Dobb (who in fact presented it very cautiously), and its opposition—emphasis on the role of towns in the rise of capitalism—is associated with Paul Sweezy. See Dobb, *Studies in the Development of Capitalism* (1947), Sweezy, "A Critique" (1976). The thesis is also central to the debates over "dependency theory." For instance, Robert Brenner argues that towns and trade were essentially irrelevant precisely because the issue is production, not exchange, and Brenner believes (wrongly) that towns were not really important points of production in the medieval world. See Brenner, "The Agrarian Roots of European Capitalism" (1985), pp. 38–39. Also see Chapter 2, note 172, above.

32. Elvin, *The Pattern of the Chinese Past* (1973).

33. Elvin, *The Pattern of the Chinese Past* (1973); Mahalingam, *Economic Life in the Vijayanagar Empire* (1951); Naqvi, *Urban Centres in Upper India, 1556–1803* (1968); Satish Chandra, "Commerce and Industry in the Medieval Period" (1964).

34. Pires, *The Suma Oriental* (1944). Also see K. N. Chaudhuri, *Trade and Civilization in the Indian Ocean* (1985); Chan-Cheung, "The Smuggling Trade Between China and Southeast Asia During the Ming Dynasty (1967); Di Meglio, "Arab Trade with Indonesia and the Malay Peninsula from the 8th to the 16th Century" (1970); Elvin, "China as a Counterfactual" (1988); Gupta, *Industrial Structure of India During Medieval Period* (1970); I. Habib, "Usury in Medieval India" (1964); Jha, *Studies in the Development of Capitalism in India* (1963); Pires, *The Suma Oriental* (1944); Prakash, "Organization of Industrial Production in Urban Centres in India During the Seventeenth Century with Special Reference to Textiles" (1964); Victor Purcell, *The Chinese in Southeast Asia*, 2nd ed. (1965); Jan Qaisar, "The Role of Brokers in Medieval India" (1974); Simkin, *The Traditional Trade of Asia* (1968); Toyoda, *History of Pre-Meiji Commerce in Japan* (1969); Udovitch, "Commercial Techniques in Early Medieval Islamic Trade" (1974).

35. Needham and collaborators, *Science and Civilization in China* (1954–1984), vol. 4, part 3; Lewis, "Maritime Skills in the Indian Ocean, 1368–1500" (1973); Lo, "China as a Sea Power" (1955); Ma Huan, *The Overall Survey of the Ocean's Shores* (1970); Purcell, *The Chinese in Southeast Asia*, 2nd ed. (1965).

36. S. Chaudhuri, "Textile Trade and Industry in Bengal Suba, 1650–1720" (1974); Elvin, "China as a Counterfactual" (1988); Gernet, *Daily Life in China on the Eve of the Mongol Invasion* (1962); Jha, *Studies in the Development of Capitalism in India* (1963); Naqvi, *Urban Centres in Upper India, 1556–1803* (1968); Needham and collaborators, *Science and Civilization in China* (1965–1984); Jan Qaisar, "The Role of Brokers in Medieval India" (1974); Rodinson, "Le Marchand Musulman" (1974); Rodinson, *Islam and Capitalism* (1973); Bodo Wiethoff, *Die Chinesische Seeverbotspolitik und der Private Überseehandel von 1368 bis 1567* (1963); Yang, "Government Control of Urban Merchants in Traditional China" (1970).

37. I proposed this idea in Blaut, "Where Was Capitalism Born?" (1976). Janet Abu-Lughod's important book *Before European Hegemony: The World System A.D. 1250–1350* (1989) is the first effort to show in precise detail how this system worked in the fourteenth century. Also see S. Chaudhuri, "Textile Trade and Industry in Bengal Suba" (1985); Simkin, *The Traditional Trade of Asia* (1968); Amin, *Accumulation on a World Scale* (1974) and *Unequal Development* (1976).

38. Steward, *Theory of Culture Change: The Methodology of Multilinear Evolution* (1955). See note 1.

39. Braudel, *The Mediterranean* (1972); Goitein, *A Mediterranean Society* (1967); Lane, *Venice: A Maritime Republic* (1973).

40. Abu-Lughod, *Before European Hegemony: The World System* A.D. *1250–1350* (1989), fig. 1, p. 34. One of the eight regions is the nonmaritime circuit extending from China through Central Asia to the Black Sea.

41. Blaut, *The National Question* (1987b).

42. Needham and collaborators, *Science and Civilization in China* (1965–1984), Vol. 5, part 7; Needham, *Gunpowder as the Fourth Power, East and West* (1985).

CHAPTER 4

After 1492

EXPLAINING 1492

In 1492, as have seen, capitalism was slowly emerging in many parts of Asia, Africa, and Europe. In that year there would have been no reason whatever to predict that capitalism would triumph in Europe, and would triumph only two centuries later.

By "the triumph of capitalism" I mean, in the present context, the political revolution that transferred power from the old feudal landlord elite to the bourgeoisie (the burghers, the capital-accumulating new elite): the bourgeois revolution. This was really a revolutionary epoch, not a single brief event, but I will follow convention by dating it to 1688, the year of England's "Glorious Revolution." In that year (minor qualifications aside) the bourgeoisie definitively took power in England. This class already held power in Holland and in some small states of southern Europe, while in some other parts of Europe (like France) the bourgeoisie was vigorously "rising" in certain regions although the conflict with feudal polities had not yet been won at the level of state power. It should be emphasized that the capitalism that triumphed was not industrial capitalism. How this preindustrial capitalism should be conceptualized is a difficult question because it is something much larger than the "simple commodity production" and "merchant capital" of earlier times. But the Industrial Revolution did not really begin until a century later, in the late eighteenth century, and those who conceptualize the Industrial Revolution as simply a continuation of the bourgeois revolution are neglecting a large and important block of history, both inside and outside of Europe.

The explanation for the rise of capitalism to political power in Europe in the (symbolic) year 1688 requires an understanding of (1) the reasons Europeans, not Africans and Asians, reached and conquered America, and thus garnered the first fruits of colonialism; (2) the reasons the conquest was successful; (3) the direct and indirect effects of the sixteenth-century plunder of American resources and exploitation of

179

American workers on the transformation of Europe; and (4) the direct and indirect effects of seventeenth-century colonial and semicolonial European enterprise in America, Africa, and Asia on the further transformation of Europe and eventually the political triumph of capitalism in the bourgeois revolution.

In the following paragraphs we will summarize each of these processes in turn, and thus, so to speak, "explain 1492." Then we will turn to the problem of explaining the rise of capitalism to political power in Europe— or more properly, part of Europe—in the period 1492–1688, in the sense of trying to sort out the significance of colonialism and the extra-European world in this epochal transformation. Finally, we will look at the significance of colonialism, and the role of non-Europe, in the initial stages of the Industrial Revolution, roughly, in the period of the late eighteenth and early nineteenth centuries, and will look at the complementary process: the beginnings of underdevelopment in Africa and Asia.

This inquiry should lead to an explanation for the fundamental fact that capitalism became centrated in Europe. I use the verb "centrate" to emphasize one crucial theoretical argument of this book: the rise of a more-or-less capitalist system had been going on in many parts of the world prior to 1492; after 1492, new forces entered which slowed, then stopped, its evolution outside of Europe and quickened it inside. Thus the rise of capitalism after 1492 was as much a matter of shifting its main headquarters to Europe as it was a matter of "rising" in a simple evolutionary sense.

Why America Was Conquered by Europeans and Not by Africans or Asians

One of the core myths of Eurocentric diffusionism concerns the discovery (so-called) of America.[1] Typically it goes something like this: Europeans, being more progressive, venturesome, achievement-oriented, and modern than Africans and Asians in the late Middle Ages, and with superior technology as well as a more advanced economy, went forth to explore and conquer the world. And so they set sail down the African coast in the middle of the fifteenth century and out across the Atlantic to America in 1492. This myth is crucial for diffusionist ideology for two reasons: it explains the modern expansion of Europe in terms of internal, immanent forces, and it permits one to acknowledge that the conquest and its aftermath (Mexican mines, West Indian plantations, North American settler colonies, and the rest) had profound significance for European history without at the same time requiring one to give any credit in that process to non-Europeans.

In reality, the Europeans were doing what everyone else was doing across the hemisphere-wide network of protocapitalist, mercantile–maritime centers, and Europeans had no special qualities or advantages, no peculiar venturesomeness, no peculiarly advanced maritime technology, and so on. What they did have was opportunity: a matter of locational advantage in the broad sense of accessibility. The point deserves to be put very strongly. If the Western Hemisphere had been more accessible, say, to South Indian centers than to European centers, then very likely India would have become the home of capitalism, the site of the bourgeois revolution, and the ruler of the world.

In the late Middle Ages long-distance oceanic voyaging was being undertaken by mercantile–maritime communities everywhere. In the fifteenth century Africans were sailing to Southeast Asia, Indians to Africa, Arabs to China, Chinese to Africa, and so on.[2] Much of this voyaging was across open ocean and much of it involved exploration. Two non-European examples are well known: Cheng Ho's voyages to India and Africa between 1417 and 1433, and an Indian voyage around the Cape of Good Hope and apparently some 2,000 miles westward into the Atlantic circa 1420.[3] In this period the radii of travel were becoming longer, as a function of the general evolution of protocapitalism, the expansion of trade, and the development of maritime technology. Maritime technology differed from region to region but no one region could be considered to have superiority in any sense implying evolutionary advantage, and novel ideas and techniques were being spread in all directions by rapid criss-cross diffusion. The entire hemisphere was participating and sharing in a Spatial Revolution.

Certainly the growth of Europe's commercial economy led to the Portuguese and Spanish voyages of discovery. But the essence of the process was a matter of catching up with Asian and African protocapitalist communities by European communities, which were at the margin of the hemisphere-wide system and were emerging from a period of downturn relative to some other parts of the system. Iberian Christian states were in conflict with Maghreb states and European merchant communities were having commercial difficulties both there and in the eastern Mediterranean. The opening of a sea route to West African gold mining regions, along a sailing route known since antiquity, and using maritime technology known to non-Europeans as well as Europeans, was an obvious strategy.[4] By the late fifteenth century the radii of travel had lengthened so that a sea route to India was found to be feasible (with piloting help from African and Indian sailors). The leap across the Atlantic in 1492 was certainly one of the great adventures of human history, but it has be seen in a context of shared technological and geographical knowledge, high potential for commercial success, and other factors that place it, in a hemispheric perspective, as something that could have been undertaken by non-Europeans just as easily as by Europeans.

Europeans had one advantage. America was vastly more accessible from Iberian ports than from any extra-European mercantile–maritime centers that had the capacity for long-distance sea voyages. Accessibility was in part a matter of sailing distance. Sofala, which is presumed to have been the southernmost major seaport in East Africa in that period (there may have been others farther south), is roughly 3,000 miles farther away from an American landfall than are the Canary Islands (Columbus's jumping-off point) and 5,000 miles farther from any densely populated coast with possibilities for trade or plunder. The distance from China to America's northwest coast was even greater, and greater still to the rich societies of Mexico.

To all of this we must add the sailing conditions on these various routes. Sailing from the Indian Ocean into the Atlantic one sails against prevailing winds. The North Pacific is somewhat stormy and winds are not reliable. From the Canaries to the West Indies, on the other hand, there blow the trade winds, and the return voyage is made northward into the westerlies. Obviously an explorer does not have this information at hand at the time of the voyage into unknown seas. The extent of the geographical knowledge possessed by Atlantic fishing communities in the fifteenth century remains, however, an unanswered and intriguing question, and there is speculation that these people fished around Newfoundland and the Grand Banks before 1492. More concretely, the Iberian sailors going to and from the Canaries, Madeira, and the Azores made use of the same basic wind circulation as did Columbus in crossing the entire ocean; Columbus knew that the trade winds (or easterlies) would assist him outbound and had good reason to believe that the westerlies would assist the return voyage. The point here is a matter of strong probabilities. Overall, it is vastly more probable that an Iberian ship would effect a passage (round trip) to America than would an African or Asian ship in the late fifteenth century, and, even if such a voyage were made by the latter, it is vastly more probable that Columbus's landfall in the West Indies would initiate historical consequences than would have been the case for an African ship reaching Brazil or a Chinese ship reaching California.

Is this environmental determinism? There is no more environmentalism here than there is in, say, some statement about the effect of oilfields on societies of the Middle East. I am asserting only the environmental conditions that support and hinder long-distance oceanic travel. In any case, if the choice were between an environmentalistic explanation and one that claimed superiority of one group over all others, as Eurocentric diffusionism does, we would certainly settle for environmentalism.

Before we leave this topic, two important questions remain to be asked. First, why did not West Africans "discover" America since they were even closer to it than the Iberians were? The answer seems to be that mercantile, protocapitalist centers in West and Central Africa were not oriented to commerce by sea (as were those of East Africa). The great long-distance trade routes led across the Sudan to the Nile and the Middle East, across the Sahara to the Maghreb and the Mediterranean, and so forth. Sea trade existed all along the western coast, but it was not large in scale, given that civilizations were mainly inland and trading partners lay northward and eastward. Second, why did the trading cities of the Maghreb fail to reach America? This region (as Ibn Khaldun noted not long before) was in a political and commercial slump. In 1492 it was under pressure from the Iberians and the Turks. Just at that historical conjuncture, this region lacked a capacity for major long-distance oceanic expeditions. Also, these cities, which traded directly with the Sudan and the gold regions, did not have the economic incentive that Europeans had to bypass the Saharan land routes in search of a new—that is, cheaper—source of gold.

Why the Conquest Was Successful

America became significant in the rise of Europe, and the rise of capitalism, soon after the first contact in 1492. Immediately a process began, and explosively enlarged, involving the destruction of American states and civilizations, the plunder of precious metals, the exploitation of labor, and the occupation of American lands by Europeans. If we are to understand the impact of all of this on Europe (and capitalism), we have to understand how it occurred and why it happened so quickly—why, in a word, the conquest was successful.

There is a second crucial reason we need to understand the causality of the conquest. A nondiffusionist history starts all causal arguments with the working assumption that Europeans had no innate superiority, in any dimension of culture, over non-Europeans, no a priori "higher potential" for progress. This leads first to a recognition that Europeans in 1492 had no special advantage over Asians and Africans, ideological, social, or material. But it demands that we make the same working hypothesis about all human communities. Why, then, did Europeans discover America instead of Americans discovering Europe (or Africa, or Asia)? And why, after the first contact, did Europeans conquer the American civilizations instead of being defeated and driven from America's shores? The working assumption of cultural uniformitarianism—or, if you prefer, the psychic unity of humankind—here confronts the diffusionist tendency to dismiss the peoples of America as primitive and irrelevant.[5]

There were several immediate reasons why American civilizations succumbed, but one of these is of paramount importance and probably constitutes a sufficient cause in and of itself. This is the massive depopulation caused by the pandemics of Eastern Hemisphere diseases that were introduced to America by Europeans.[6] A second factor was the considerable advantage Europeans held in military technology, but this advantage has to be kept in perspective. The technological gap was not so great that it could by itself bring military victory—after the initial battles—against American armies that were vastly larger and would sooner or later have adopted the enemy's technology. America is a vast territory, and in 1492 it had a very large population, numbering at least 50 million people and conceivably as many as 200 million, a goodly proportion of these people living in state-organized societies with significant military capability.[7] Military technology tends, historically, to diffuse from one camp to the opposing camp in a relatively short time. Moreover, the superiority of the Spaniards' primitive guns was not really very great when compared with the Americans' bows and arrows. I think it is, therefore, certain that the tide would have turned against the Europeans had the matter been merely one of military capability. There would have been no conquest, or the conquest would have embraced only a limited territory, and certainly would not have swept south as far as the great civilizations of the central Andes. The point is that history went in a different direction because of the incredibly severe and incredibly rapid impact of introduced diseases. Resistance collapsed because the Americans were dying in epidemics even before the battles were joined.[8] Probably 90% of the population of central Mexico was wiped out during the sixteenth century; the majority of these deaths occurred early enough to assist the political conquest. Parallel processes took place in other parts of the hemisphere, especially where there were major concentrations of population, these in most cases being areas of state organization and high civilization. Perhaps three-quarters of the entire population of America was wiped out during that century.[9] Millions died in battle with the Spaniards and Portuguese and in forced-labor centers such as the mines of Mexico and Peru, but much greater numbers died in epidemics, and this was the reason that resistance to the conquest was rapidly overcome in most areas.

Both the susceptibility of American populations to Eastern Hemisphere diseases and the lower level of military technology among Western Hemisphere peoples can be explained in fairly straightforward cultural-evolutionary terms, although evidence bearing on the matter is partly indirect. The Western Hemisphere was not occupied by humans until very late in the Paleolithic period; there is dispute about the first arrivals, but most scholars do not believe that the Americas were occupied before 30,000 B.P. The first immigrants did not possess agriculture. The earliest migrations preceded the Agricultural Revolution in the Eastern Hemisphere; in addition, the source area for the migrations, northeastern Siberia, is generally too cold for agriculture, even for present-day agriculture, and we would not expect to find that these cultures were experimenting with incipient agriculture 20,000 years or so ago although some low-latitude cultures were doing so. Migrants to America were paleolithic hunters, gatherers, fishers, and shellfishers. They came in small numbers, apparently in a widely spaced series of relatively small population movements, and spread throughout both North and South America. Only after some millennia had passed was the stock of resources for hunting, fishing, gathering, and shellfishing under any significant pressure from humans. One assumes that population growth was slow but—this is of course speculative—that population growth eventually did reach the point where conditions were favorable to an Agricultural Revolution.[10] In the Eastern Hemisphere the Agricultural Revolution seems to have occurred (as a qualitative change) roughly 10,000–12,000 years ago. In the Western Hemisphere that point may have been reached about 4,000 years later.[11] Thereafter, cultural evolution in the Western Hemisphere proceeded along lines somewhat parallel to those of Eastern Hemisphere evolution: the development of agricultural societies, of monumental ceremonial centers, science, writing, cities, feudal class structures, and mercantile trade. It seems, indeed, that the Western Hemisphere societies were closing the gap. But in 1492, military technology in the most advanced and powerful states was still well behind that of Eastern Hemisphere states. Metal was just coming into use in this arena, and guns had not been invented. Hence the superiority of Cortés's armies over Moctezuma's and Pizarro's over the Incas'. (When Cortés arrived at Tenochtitlán the Aztecs were already dying in great numbers from European diseases which, apparently, had been carried by American traders from Cuba to Mexico. Likewise, the Incas apparently were succumbing to these diseases before Pizarro arrived.[12])

The susceptibility of American populations to Eastern Hemisphere diseases, and the consequent devastation of American settlements, collapse of states, and defeat and subjugation by the Europeans, is explained within the same general model. Small populations entered America and probably bore with them only a small subset of the diseases that existed in the Eastern Hemisphere at the time of their departure. They came, in addition, from a rather isolated, thinly populated part of the hemisphere, and a part which, having a cold climate, would have lacked some diseases characteristic of warm regions. Perhaps more important is the history of the diseases themselves. Many diseases originated or became epidemiologically significant during or after the Agricultural Revolution, and have ecological connections to agriculture, to urbanization, to zoological and botanical changes in the ecosystems strongly modified by human land use, and so on. In the Eastern Hemisphere humanity entered these ecological situations after the initial migrations to the Western Hemisphere, hence these migrants to America would not have carried these diseases with them. Later migrants may have done so (although this is again unlikely because they came from a cold and isolated part of Asia, and came in small numbers). But we can assume that the sparse settlement, the hunting-gathering-fishing-shellfishing way of life, and the absence of agricultural settlements and urbanization in the Americas during many millennia, would have caused a disappearance of some of the Eastern Hemisphere diseases that had been carried across to the Americas by migrants. After a time the American populations would have lost their physiological immunities to diseases no longer present in these populations, and they would of course lack immunities to diseases never before encountered. It is known, in this regard, that utter devastation was produced in the Americas from diseases to which Eastern Hemisphere populations had such high levels of immunity that they experienced these diseases as minor maladies only.

Hence there is no need to take seriously any longer the various myths that explain the defeat of the Americans in terms of imputed irrationality or superstitiousness or any of the other classical, often racist, myths about American civilizations in 1492. (The most widely known of these myths is the idea that Mexicans imagined that Cortés and his troops were gods, and fell down before them in awe instead of fighting. This did not happen.) The relatively minor difference in technology between the two communities, and the impact of Eastern Hemisphere diseases upon Western Hemisphere communities, can be explained in terms of the settlement history of the Western Hemisphere and its consequences. The Americans were not conquered: they were infected.

Europe in 1492

In 1492, European society was rather sluggishly moving out of feudalism and toward capitalism. Nothing in the landscape would suggest that a revolutionary transformation was imminent, or even that the social and economic changes taking place were very rapid. The growth of the

English woolen trade in the fifteenth century was not (as it is often depicted) a sign of revolutionary economic change: it was complemented by a decline in competing woolen industries in southern Europe.[13] Rural growth in this century reflected mainly population recovery (in some areas) after the great plagues of the preceding century, and the commercialization of agriculture that was then taking place had been doing so for some time.[14] Towns were growing, but only slowly, and the urban population was still only a small fraction of total population (except in Italy and the Low Countries); and the urban population of Europe was smaller than it was in many non-European areas.[15] There were strong signs even of economic contraction instead of growth.[16] The Italian Renaissance, in economic terms, did not raise the Italian centers above the level of many non-European centers, including those in nearby Islamic countries (for instance, Cairo), and the Renaissance was not at all a technological revolution.[17] All of this needs to be said by way of setting the stage. Before 1492 there was slow growth in Europe, perhaps even a downturn. Certainly—and this is accepted by the majority of European historians—no truly revolutionary transformation was underway in 1492.

Within a few decades after 1492 the rate of growth and change speeded up dramatically, and Europe clearly entered a period of rapid metamorphosis. There is no dispute about this fact, which is seen in the known statistics relating to prices, urban growth, and much more beside.[18] What is disputable is the causal connection between these explosive sixteenth-century changes and the beginnings of economic exploitation in America (and, significantly but secondarily, in Africa and Asia). There is agreement that the effect was profound. But did it truly generate a qualitative transformation in Europe's economy? Or did it suddenly quicken a process already well underway? Or did it merely modify this process slightly with effects such as inflation? This question cannot be answered unless we break out of the European historical tunnel and look at what was going on in America, Asia, and Africa between 1492 and 1688, the symbolic date for Europe's bourgeois revolution.

COLONIALISM AND THE RISE OF EUROPE, 1492–1688

Colonialism and Capitalism in the Sixteenth Century

Enterprise in the Americas was from the start a matter of capital accumulation: of profit. No matter if some elements of medieval law were incorporated in legal and land-granting systems in (for

Europeans) the New World, and if the Iberian governments took a substantial, though usually overestimated, portion of the profits. The goal of all European individuals and groups involved in the enterprise, clergy apart, was to make money, for oneself or one's country (usually the former).

The leading group, almost everywhere, was the European protocapitalist class, not only merchants but also industrialists and profit-oriented landlords—not only Iberian but also Italian, Flemish, Dutch, German, English, and so on. This class community took its profit from American enterprise and invested part of it in Europe, buying land and developing commercial agriculture, developing industries (like shipbuilding, sugar refining, and so on) that were associated with the growing colonial enterprise, developing profitable businesses in spheres of activity which served the growing European economy (for instance, the burgeoning Atlantic fisheries), building urban structures, and the like. Part of the profit was plowed back into additional colonial risk enterprise, in America and in the new trading enterprises in southern Asia, Africa, and the Levant. One of the deceptively subtle aspects of the process was the immense increase in purchases of all sorts by European merchants in all markets, inside and outside of Europe, growing out of the fact that these merchants now had incredible amounts of precious metals or metal-based money at their command and could offer previously unheard-of prices. Perhaps half the gold and silver brought back from America in the sixteenth century was contraband, hence available directly for this kind of enterprise, but the remainder, after passing through the great customs houses, quickly entered circulation as the Iberian governments paid out gold and silver for goods and services.[19]

Colonial enterprise in the sixteenth century produced capital in a number of ways. One was gold and silver mining. A second was plantation agriculture, principally in Brazil. A third was the trade with Asia in spices, cloth, and the like. A fourth and by no means minor element was the profit returned to European investors from a variety of productive and commercial enterprises in the Americas, including profit on production for local use in Mexico, Peru, and elsewhere; profit on sale of goods imported from Europe; profit on a variety of secondary exports from America (leather, dyestuffs, etc.); profit on land sales in America; profit returned to Europe by families and corporations holding land grants in Mexico and other areas. A fifth was slaving. A sixth, piracy. Notice that all of this is normal capital accumulation; none of it is the mysterious thing called "primitive accumulation." [20] (Value from wage labor, not to mention forced labor, was involved, and much of it was value from production, not simply from trade.) Accumulation from these sources was massive. It was massive enough so that the process cannot be dismissed as a minor adjunct of protocapitalist accumulation in Europe itself, and it was massive enough, I believe, to fuel a major transformation in Europe, the rise to power of the bourgeoisie and the immense efflorescence of preindustrial capitalism, in ways that we will discuss.

Precious Metals

We notice first the export of gold and silver from the Americas and its insertion within the circuits of an Eastern Hemispheric market economy in which gold and silver already provide the common measure of value, directly or indirectly, in almost all markets. The flow of precious metals began immediately after the European discovery of America, and by 1640 at least 180 tons of gold and 17,000 tons of silver are known to have reached Europe.[21] (The real figures must be at least double these amounts, since records were poor for some areas and periods and since contraband was immensely important.[22]) Additional quantities of gold came from colonial activities in Africa. In the period 1561 to 1580 about 85% of the entire world's production of silver came from the Americas. The simple quantity of gold and silver in circulation in the Eastern Hemisphere economy as a whole was profoundly affected: hemispheric silver stock may have been tripled and gold stock increased by 20% during the course of the sixteenth century as a result of bullion brought from America.[23] The fact that much of the pre-existing stock must have been frozen in uses not permitting direct or indirect conversion to money suggests to me that American bullion may have as much as doubled the gold- and silver-based money supply of the Eastern Hemisphere as a whole. (In Europe, the circulation of metal coins increased eight- or ten-fold in the course of the century.[24]) This process must be seen in perspective: it is money flowing constantly and in massive amounts into Europe, through Europe, and from Europe to Asia and Africa, constantly replenished at the entry points (Seville, Antwerp, Genoa, etc.) with more American supplies, and constantly permitting those who hold it to offer better prices for all goods—as well as labor and land—in all markets, than anyone else had ever been able to offer in prior times.

The importance of these flows of gold and silver is routinely underestimated by scholars, mainly for three reasons (apart from implicit diffusionism, the simple tendency to undervalue causal events in non-Europe). First, the process is seen somehow as purely primitive accumulation. But the metals were mined by workers and transported by workers; the enterprise overall involved risk capital and all of the other familiar traits of the sorts of protocapitalist productive enterprises which were characteristic of that time (that it was partly state-controlled does not alter this argument, nor does the fact that some of the labor was unfree); and very major economic and social systems were built around the mines themselves in Mexico, Peru, and other parts of America.

Second, the argument that precious metal flows significantly affected the European economy is dismissed by some scholars as "monetarism" (roughly, the theory that changes in money alone are very significant for changes in the economy overall). The error in this charge is a failure to see the sixteenth-century economy in its own, appropriate, geographical and social context, and to impute to the economy of that time the liquidity of exchange and the relative lack of spatial friction that characterizes the capitalist economy of our own time. Two facts here are basic. First of all, the possession of precious metal was highly localized in space. European merchants, as a community, obtained it and set it in motion outward, toward rural Europe and toward markets outside of Europe. Second of all, the supply of precious metal was essentially continuous, and therefore the advantage held by European protocapitalists in terms of prices they could offer for commodities, labor, and land was persistently higher than the prices which competitors anywhere could offer. So the protocapitalist community very steadily undermined the competition in all markets across the Eastern Hemisphere, within Europe and without, eventually gaining control of most international seaborne trade in most of the mercantile–maritime centers from Sofala to Calicut to Malacca.[25] The penetration of these markets, the acquisition of trading bases, and the control of a few small but important producing areas (like some islands of the Moluccas), was not a matter of European rationality or venturesomeness, but rather reflected the availability to Europeans of American gold and silver, trans-shipped through Lisbon, Antwerp, Acapulco (in the "Manila galleons"), and so on.

A third sort of doubt about the importance of American gold and silver is associated with the critique of Earl Hamilton's classic theory that the precious metal supply produced an imbalance between factors of production in the European economy, produced thereby a windfall of profits, and thus in effect destabilized the economy and moved it toward capitalism.[26] Hamilton was one of the few economic historians to perceive that American gold and silver was a crucial, central cause of change in Europe, although he was (partly) wrong about the mechanisms that brought about this change. The metals did not transform the economy in any direct sense. Rather, they enriched the protocapitalist class and thereby gave them the power to immensely accelerate the transformation that was already underway—not only in Europe—toward capitalism as a political and social system, and to prevent non-European capitalists from sharing in the process. American bullion hastened the rise of capitalism and was crucial in the process by which it became centrated in Europe.

Plantations

The impact of the slave plantation system on Europe's economy was felt mainly in the seventeenth century and thereafter. But part of the general undervaluing of the significance of early colonialism—of the world outside of Europe—is a tendency not to notice that the plantation system was of considerable importance even in the sixteenth century. Moreover, the early history of the Atlantic sugar plantation economy gives a revealing picture of the way in which the protocapitalist colonial economy was eroding the feudal economy. Sugar planting was not a new enterprise; sugar (contrary to myth) was not a rare commodity, and sugar planting (also contrary to myth) was not an insignificant economic curiosity at the fringe of capitalist development. Commercial and feudal cane sugar production was found throughout the Mediterranean in the fifteenth century.[27] Although little is known about the way planting was organized, it is known that commercial sugar production was important in India 2,000 years ago (apparently it was a Mauryan state industry), and in the Middle Ages commercial sugar planting under various feudal and probably protocapitalist systems of organization was found in East Africa, part of West Africa, Morocco, Egypt, Cyprus, the Levant, various parts of Mediterranean Europe, and other regions.[28] If cane sugar was not an important commodity in northern Europe this was because of its price, as against that of sweeteners like honey. Europeans first moved the commercial plantation system out into the newly settled Atlantic islands from Madeira to São Tomé and then vastly expanded production in the Americas. But throughout the sixteenth century the new plantations merely *supplanted* the older Mediterranean sugar-producing regions; total production for the Europe–Mediterranean market did not rise until later.[29] This was capitalist production displacing feudal and semifeudal plantation production, using the twin advantages of colonialism: empty land and cheap labor. No other industry was as significant as the plantation system for the rise of capitalism before the nineteenth century.

In 1600 Brazil exported about 30,000 tons of sugar with a gross sale value of £2,000,000. This is about double the total annual value of *all* exports from England to *all* of the world in that period.[30] It will be recalled that British exports in that period, principally of wool, are sometimes considered paradigmatic for the "awakening," indeed the "rise," of early-modern Europe. Also in 1600, per capita earnings from sugar in Brazil, for all except Indians, was about equal to per capita income in Britain later in that century.[31] The rate of accumulation in the Brazilian plantation industry was so high at the end of the sixteenth century that it was able to generate enough capital to finance a doubling of its capacity every two years.[32] Early in the seventeenth century the Dutch protocapitalist community (which was heavily involved in the Brazilian sugar enterprise, mainly in the shipping and sales dimensions) calculated that profit rates in the industry were 56% per year, totalling nearly £1,000,000 per year. The rate of profit was higher still a bit earlier, at the close of the sixteenth century, when production costs, including the cost of purchasing slaves, amounted to only one-fifth of income from sugar sales. These statistics should be seen against the background of an industry that was not responding to some novel demand for some novel product in an already-rising Europe, but was merely (in essence) undercutting the precapitalist Mediterranean producers of Spain, Italy, Morocco, Egypt, and elsewhere, in the supply of a highly important commercial product.

Sugar is of course the centerpiece of the plantation system down to the late eighteenth century. But other kinds of colonial production, mainly but not only agricultural, and fully as close to capitalism as was the Brazil plantation system, were of some significance even before the end of the sixteenth century. There was, for instance, some direct production of spices in the Moluccas and some European involvement with Indian merchant capitalists in the organization pepper production in South India. Dyes, tobacco, and other commercially valuable products were flowing from America to Europe. A very large agricultural economy existed in parts of America to supply food, fiber, leather, and other necessities to the mining settlements and other settlements. Immediately after 1492 (or before?) west European fishermen and whalers developed an immense industry in Newfoundland and elsewhere on the North American coast.

To all of this must be added the profits from other sorts of colonial and semicolonial activities in the Eastern Hemisphere.[33] The slave trade was highly profitable even in the sixteenth century. European merchant capitalists of all nations profited greatly from the Lisbon trade with Asia and East Africa in textiles and particularly spices (the Asian spices carried by the Portuguese and sold mainly through Antwerp did not replace the traditional Mediterranean flow but rather added to it, hence providing a novel and important source of accumulation). There was, in addition, considerable profit from the within-Asia trade resulting from the domination of long-distance oceanic trade in East Africa, India, and

Southeast Asia by Portugal (with participation also by Spain and later Holland). Broadly speaking, however, accumulation deriving from Western Hemisphere colonial activities far outweighed that from Eastern Hemisphere activities, colonial and semicolonial, in the sixteenth century. Overall, both the quantitative significance, in that century, of production and trade in colonial and semicolonial areas and the immense profitability of the enterprise, that is, the rapid capital accumulation which it fostered directly and (in Europe) indirectly, add up to a significant vector force, easily able to change the process of economic transformation in Europe from sluggish evolution to rapid revolution.

Effects

There seem to be two particularly good ways to assess the real significance for the rise of capitalism of sixteenth-century colonial production in America, and some other areas, along with trading, piracy, and the like, in Asia and Africa. One way is to trace the direct and indirect effects of colonialism on European society, looking for movements of goods and capital, tracing labor flows into industries and regions stimulated or created by colonial enterprise, and looking at the way urbanization flourished in those cities that were engaged in colonial (and more generally extra-European) enterprise or closely connected to it, and the like. This processes overall would then be examined in relation to the totality of changes taking place in Europe in that century, to determine whether, in Europe itself, changes clearly resulting from the direct and indirect impact of extra-European activities were the prime movers for economic and social change. This task still remains undone. The second way is to attempt to arrive at a global calculation of the amount of labor (free and unfree) that was employed in European enterprises in America, Africa, and Asia, along with the amount of labor in Europe itself which was employed in activities derived from extra-European enterprise, and then to look at these quantities in relation to the total labor market in Europe for economic activities that can be thought of as connected to the rise of capitalism. This task has not been done either; indeed, as far as I know little research has been done on sixteenth-century labor forces and labor markets in American settlements or indeed in Europe. So the proposition which I am arguing here, concerning the significance of sixteenth-century colonialism (and related extra-European activities) for the rise of capitalism in Europe, perhaps cannot be tested as yet.

Still, there are suggestive indications. Some of these have been mentioned already: matters of assessing the quantities and values of colonial exports to Europe. We can also speculate about labor. One approach is through population. The population of Spain and Portugal in the mid-sixteenth century may have been around nine million.[34] Estimates of sixteenth-century populations for America vary widely and there is much controversy about population levels and rates of decline,[35] but for the present, highly speculative, and essentially methodological, argument, I will ignore the controversies and play with global estimates. The population of Mexico at midcentury may have been around six million, a population that was undergoing continuous decline from its preconquest level of perhaps 30 million down to one-tenth of that figure (or perhaps less) in 1600.[36] Populations in the Andean regions involved in mineral and textile production for the Spaniards may (I am speculating) have totalled five million in the late sixteenth century. Perhaps we can add an additional two million for the population of other parts of Ibero-America that were within regions of European control and presumably involved, more or less, in the European-dominated economy. Let us, then, use a ball-park estimate of 13 million for the American population that was potentially yielding surplus value to Europeans in the mid-to-late sixteenth century. The population seems larger than that of Iberia. Granted, the comparison should be made with a larger part of Europe, certainly including the Low Countries, which were intimately involved in the exploitation of America (and Asia) at this period, along with parts of Italy and other countries. Assume then a figure of 20 million for Europe as against 13 million for America.

I see no reason to argue that the European populations were more intimately involved in the rise of capitalism than the American populations—that is, the 13 million people who we assume were in European-dominated regions. It is likely that the proportion of the American population that was engaged in labor for Europeans, as wage work, as forced labor including slave labor, and as the labor of farmers delivering goods as tribute or rent in kind, was no lower than the proportion of Iberian people engaged in labor for commercialized sectors of the Spanish and Portuguese economy. Moreover, the level of exploitation for Indian labor must have been much higher than that for Iberian labor because portions of the Indian labor force were worked literally to death in this period—depopulation was due in part to forced labor—and so the capital generated by each American worker must have been higher than that generated by a European worker. (We need to remind ourselves again that we are dealing with a preindustrial, basically medieval economy in Europe. It cannot be argued, for instance, that technology or fixed capital in production was more advanced in the utilization of European than in that of American labor, so exploitation was in the last analysis a function of human effort.)

We must next take into account the fact that the capital accumulated from the labor of Americans went directly to the economic sectors in Europe that were building capitalism, whereas most workers and peasants in Europe were still connected to essentially medieval sectors of the economy. Then we must add the labor of Africans and Asians. And finally, we must take into account the European workers, in Europe and elsewhere, whose labor must be considered part of the extra-European economy. By this admittedly speculative reasoning, free and unfree workers in the colonial and semicolonial economy of the late sixteenth century were providing as much or more surplus value and accumulated capital for European protocapitalism, the rising bourgeoisie, as were the workers of Europe itself.

Little is known about the American work force in the sixteenth century, but, again, some speculations are possible. Las Casas asserted that three million or more Indians had been enslaved by the Spaniards in the northern part of Spanish America during the first half of the sixteenth century, and this figure, once dismissed, is now taken seriously.[37] It is known that more than 400,000 were enslaved in Nicaragua alone.[38] It is realized also that Indian slave labor was extremely important in the European economy of America in that period, in Brazilian sugar planting, Mesoamerican and Antillean mining, and elsewhere. Let us speculate that 100,000 Indians were working as slaves for Spaniards in a given year in the mid-sixteenth century. Perhaps 20,000 Indians were working at free and forced labor in the mines of Mexico and the Andes in the latter part of the century,[39] and it is safe to assume that five times that number were involved in the mining economy overall. Potosí, the great Andean silver-mining city, had a population of 120,000 in the 1570s (larger than Paris, Rome, Madrid, Seville). A much greater but unknown number of Indians were workers on haciendas and other European enterprises, or provided periodic forced labor, or provided tribute and rent in kind. (The Cortés *encomienda* in Mexico included 50,000 Indians.[40]) There may have been 100,000 African slaves in America and on the island of São Tomé in the closing years of the century.[41] There may have been 300,000 Europeans, Mulattos, and Mestizos in the Americas in 1570,[42] of whom conceivably as many as 250,000 were workers.

Perhaps it would not be unreasonable to estimate that one million people were working in the European economy of the Western Hemisphere in the closing years of the sixteenth century, perhaps half of them engaged in productive labor in distinctly capitalist enterprises. Can this have been more than the European protocapitalist work force of the time? All of this is somewhat speculative, but it points toward the conclusion that American labor was a truly massive part of the total labor of this we must add three additional quantities: labor involved in the slave trade within continental Africa;[43] labor in other extra-European regions (São Tomé, Ternate, Calicut, and so on) that was incorporated into the European economy or produced goods for trade to Europeans; and labor of Europeans, inside and outside of Europe, which was part of the extra-European economy—sailors, soldiers, stevedores, teamsters, clerks, foremen, and the rest.

By the end of the sixteenth century the rise of Europe had well begun. As capital flowed into Europe, and as other effects of colonial enterprise also flowed into the European system or region, secondary causation appeared, including agricultural expansion and transformation, primitive manufacturing, urbanization, and expansion of rural settlements and the commercial economy. These latter have been looked at carefully but in a mainly tunnel-historical framework; as a result, the rise of Europe in the sixteenth century has appeared to be a process taking place wholly within the European spatial system, and caused wholly (or mainly) by autochthonous forces. As we have seen, this is an inaccurate picture and an incomplete one. Urbanization was taking place, but mainly in areas connected to the extra-European economy. Inflation was also (with some qualifications) most severe in these areas.[44] Among the sectors of the European economy that were growing in the sixteenth century, some, like piracy and shipbuilding, were tied directly to the extra-European economy, while others, like wheat production and North Atlantic fishing, were stimulated directly and indirectly by that economy.[45]

I would generalize as follows. The initiating condition, at the beginning of the sixteenth century, is a west and central European economy that is undergoing slow but definite change toward capitalism—as are many regions of Asia and Africa at that same time. Novel forces intrude into the European system, as impinging boundary processes, because of the conquest of America and the other extra-European events, intruding processes which consist mainly of capital and material products (and of course the labor embedded in these things). These then intersect with the ongoing evolving economic, technological, demographic, and other changes. Many new changes appear, as a result not of direct stimulus from the extra-European world but from the changes already underway, which themselves are mainly results of those extra-European boundary processes. The internal European changes of course feed forward to produce intensification of the processes going on in America, Asia, etc., and these, in turn, produce still more changes within Europe.

We can see a geographical pattern in all of this. There is a tendency for major economic changes to occur first near the mercantile–maritime centers that participate in the extra-European processes. Obviously, not centers that participate in the extra-European processes. Obviously, not all of the centers that existed in 1492 were equal participants in that process, with some of the Iberian, Italian, and Flemish port cities taking the lead. But the network was sufficiently tight so that Hanseatic and English ports were early participants, as were inland cities with special economic characteristics, like Augsburg and Paris. From these many centers, the process spread into the interior of Europe, first into areas that supplied basic staple goods like wheat—the growth at that time of the Baltic wheat trade and manorial production of wheat in parts of central and east Europe is well known—and then elsewhere. At any given time we see a broad and irregular spatial pattern (of the type which geographers call "distance decay") of descending levels of urbanization and commercial production as we move across the landscape toward interior Europe.

Other processes were underway as well, and so the pattern that I have mapped out here is much too simple. Population growth in some areas reflected sixteenth-century economic changes associated with extra-European events but in other areas it signified recovery from the fourteenth- and fifteenth-century population declines. Other changes, such as peasant revolts, reflected the general crisis of the late feudal economy, but the sixteenth-century rise of prices and (at least in some areas) rents was a contributing force in the unrest. As to the Reformation, I would argue in the Tawneyan tradition that it was broadly an effect, not an independent cause, of the economic changes that were taking place in Europe in the sixteenth century.[46] But which changes? The internal crumbling of feudalism? The forces impinging from the extra-European world? Both? Probably the spatial diffusion of the Reformation in the sixteenth century reflected mainly intra-European forces,[47] but by the time of the seventeenth-century bourgeois revolutions, the areas most deeply involved in extra-European activities tended to be centers also of Protestantism. In short: the spatial patterns of change in sixteenth-century Europe reflect to some extent the integration of Europe with America, and secondarily Africa and Asia, but the pattern is still somewhat unclear.

Overall, the processes of transformation and modernization in sixteenth-century Europe were terribly complex, varying in time and place throughout most of that continent. But the generalization is nonetheless fairly straightforward. The extra-European component, after 1492, led to an immense stimulation of changes in Europe, those that produced on the one hand an increase in the rate of European economic change and growth, and on the other hand the beginnings of a centration of capitalism in Europe (a process discussed further below). By the end of the sixteenth century these extra-European forces had laid the foundation for the political and social triumph of (preindustrial) capitalism, or rather for the fact that the Glorious Revolution occurred in 1688, instead of much later, and in England, instead of Egypt or Zimbabwe or India or China (or all of these at once).

Colonialism and Capitalism in the Seventeenth Century

By the middle of the seventeenth century, changes were taking place in Europe at a rapid rate and on a massive scale, and the problem of sorting out the internal and external causes and effects for this period is a very complex matter. In this same period there occurred a massive expansion, in location and intensity, of formal and informal colonialism in the Americas and around the coasts of Africa and Asia, and for these extra-European processes the problem of complexity is compounded by a lack of quantitative data regarding volume of production, numbers in the labor forces, capital accumulation, and other information that would help us to judge the role of colonialism (as a broad concept) in the changes that were taking place within Europe. These matters are far too complex to permit us to discuss them satisfactorily here. I will limit myself to a rather sketchy intervention or (if you prefer to call it that) a model.

By the beginning of the seventeenth century, the Netherlands and England had emerged as the centers (or center) of capitalist development in Europe.[48] Although Spain continued to feed huge quantities of silver and some gold into Europe in the first half of this century, and Portuguese plantations in Brazil and trading activities in Asia continued to be important fountains of accumulation, the main expansion of colonial enterprise after 1600 was Dutch and English. The crucial component was the West Indian plantation system, which expanded explosively after about 1640. (Fifty thousand slaves were imported into Barbados alone in the following 50 years. Probably two *million* slaves were imported into America in the course of the seventeenth century.)[49] If we place the Dutch and British sugar colonies in the same economic space as the metropolitan countries themselves, it seems likely that the sugar plantation economy was the single largest productive sector in this expanded European economy (or "Atlantic economy," as it is often called) aside from family farming, and by far the largest single generator of value. (Brazilian plantations, producing partly for Dutch capital, were still, in the mid-seventeenth century, more massive than the West Indian.) But British and Dutch enterprise in the Eastern Hemisphere was also expanding very rapidly; the East India companies were formed around 1600, and by 1650 the Dutch and British together controlled most of the intercontinental trade—unequal trade, and in a sense semicolonial trade—with Asia, as well as the slave trade in Africa. Meanwhile, Spanish enterprise was yielding substantial accumulation in America (whether or not there was a "seventeenth-century depression"). And we must not ignore the great variety of additional extra-European sources of accumulation: a now massive fishing industry in the northwestern Atlantic, resource extraction and the beginnings of European settlement in North America, the slave trade, piracy, Russian enterprise in Siberia, and much more.

The key question is this: How central was the role played by colonial and semicolonial enterprise in the seventeenth-century rise of Europe and of capitalism within Europe? The model I would build involves two elements. The first is a continuation and enlargement of the sixteenth-century processes, which, as I argued, involved a sluggishly growing European economy quickened into rapid development by extra-European forces after 1492. By the middle of the next century the European bourgeoisie had strengthened their class position and (in the key locations) had enticed much of the feudal aristocracy into joining bourgeois enterprise,[50] and had well begun the process of destroying protocapitalist enterprise outside of Europe, as a result of the inflowing capital from America (and secondarily in that period, Africa and Asia).

Now, apart from stocks of precious metal, it is improbable that capital accumulated from extra-European enterprise in 1500–1650 amounted to a sizeable share of total invested capital in Europe, even in the more advanced regions of Europe, even in the economic sectors in which capital was more or less fluid. What it did do was provide a critical increment: everywhere it allowed the merchant-entrepreneurial community to offer higher prices for products, labor, and land; everywhere it put investment capital in the hands of classes and communities other than the traditional elite, the group less likely to accumulate beyond its social needs and less likely to reinvest profits in new ventures. Colonial capital, in a word, was new capital. Without it, the sluggish late-medieval economy of pre-1492 days would have continued its slow progress out of feudalism and toward capitalism (or something like capitalism), but there would have been no Seventeenth-Century Bourgeois Revolution.

Perhaps the essence of capitalism, at a level of aggregation above the worker–capitalist class relation, is the reinvestment of profits to enlarge productive capacity. Capitalist enterprise can be technologically primitive or advanced but always, to survive, it must accumulate capital. It is never in equilibrium. This point leads us to focus on the conditions that permitted continued growth, exponential growth, in the sixteenth and seventeenth centuries. This growth did not involve technological change in any important way: production increases were mainly matters of drawing more workers and more productive materials into traditional productive processes to yield more output. Given the fact that capital for expansion was available because of the extra-European enterprises and other, related developments, the key problem in the seventeenth century must have been markets, or demand. The capitalist had access to capital, had access to labor—at the levels of production then prevailing a truly massive proletarianization was not necessary—and had access to raw materials (some European, some colonial). The growth of a capitalist enterprise in that period was perhaps constrained most seriously by the need to open up new markets: to sell more of the product so that more could be produced, more capital generated, and so on.

Some of these markets were in Europe itself, reflecting at first the ability of capitalist enterprise to sell traditional products (like sugar) at lower costs than prevailed under the feudal economy, but gradually the urbanization and commercialization of the continent brought in feedback loops so that the newer way of life, generated by the rise of capitalism, itself generated more internal markets for capitalism. But probably the main growth of markets for protocapitalist enterprise in the seventeenth century, and thus the main stimulus for the rise of capitalism, was *outside* the system. This is well known in the case of trade with eastern Europe. It is known in the case of markets in America, Africa, and Asia, but the quantitative significance of these extra-European markets has not been fully evaluated. In the case of the English bourgeoisie, the main markets for capitalist enterprise, including agricultural and nonagricultural products from England and re-exported products from abroad, were in America, Africa, and Asia, along with nontraditional markets in the Baltic. For the Dutch, extra-European commerce was even more important. Italian communities continued to depend considerably on the eastern Mediterranean.

In the seventeenth century, then, the crucial role of the extra-European world, added on to and perhaps more significant than its sixteenth-century role as provider of bullion and some other products, was to permit an expansion of demand—including forced demand, as on the slave plantations—for capitalist products, a demand sufficiently great so that productive capacity and output of capitalist enterprises could continue to grow at an incredibly fast rate. This growth in output was one of the two essential seventeenth-century forces involved in the rise of capitalism. The second force was, simply, the political triumph itself, the bourgeois revolution. This provided the bourgeoisie with the legal and political power to rip apart the fabric of the society in its quest for accumulation. Forced proletarianization thereby became possible, as did government support for almost any strategy that the new accumulating elite had in mind. And an Industrial Revolution, a transformation of the methods of production so that output could increase at an even greater rate, became (one might say) inevitable.

THE CENTRATION OF CAPITALISM

The phrase "the rise of capitalism" generally evokes an image of factories, steam engines, masses of wage workers, cities grimy with coal dust: *industrial* capitalism. Our discussion thus far has not dealt with the rise of industrial capitalism—the Industrial Revolution—but with the precursors to that momentous event. But let me, for a moment, review some of these precursors.

Before 1492, most of the preconditions that would be critical for the eventual rise of industrial capitalism were present not merely in parts of Europe but also in parts of Asia and Africa. After 1492, in the sixteenth and seventeenth centuries, Europe acquired three additional preconditions. One was the very considerable accumulation of wealth from the mines and plantations of America and from trade in Asia and Africa. The second, closely related to the first, was the huge enlargement of markets outside of western Europe for products either produced in western Europe or imported and then reexported; that is, a very great and almost constantly growing demand. Third, and most important of all, the social sectors involved with capitalism took political power on a wide scale in western Europe, something that had not happened elsewhere except on very small terrains. This, the bourgeois revolution, allowed the emerging capitalist class-community to mobilize state power toward its further rise, such that the entire society contributed to the underwriting of colonial adventures and to the preparation of infrastructure such as cities and roads, while the state's police and military power could now be mobilized to force people off the land and into wage work, and to conscript people and resources for advantageous wars abroad. All three of these precursors, as I have argued, appeared because of—or would not have appeared had it not been for—colonialism.

Historians engage in fierce debates about the causes of the Industrial Revolution. Most of the candidate causes, or "factors," are theories within the "European miracle" category which we discussed and tried to refute in Chapter 2. Propositions about, for instance, general medieval modernization of the European economy and polity, medieval technological revolutions, "rationality" in medieval and later times, and the like, are built into the most common explanations for the later emergence of an Industrial Revolution. We showed, I hope, that all of these processes were at work outside of Europe as well as inside, so that they cannot be enlisted as causes of an event that happened only in Europe.

This is a problem where the sequence and dating of events is extremely important. The concept of the Industrial Revolution is usually bound up with two more specific transformations: the development of steam power and generally novel technology in industrial production, and the development of wage labor in industrial production. But the timing is wrong. The technological part of the Industrial Revolution became important very late in the process, too late to explain the revolution itself. It is certainly true that technological advances were taking place in European manufacturing during the period from 1492 to, say, 1750, but very little of this technology was unique to Europe, as we have seen, and, most crucially, the technological advances that eventually became important in increasing manufacturing production and increasing labor efficiency in that production occurred much later: in the last decades of the eighteenth century and, much more profusely, in the nineteenth century. In agriculture, the main technological advances were matters of increasing areal productivity in an environment of declining agricultural labor, but all of the essential technological changes that were involved in this process were traditional and were known outside of Europe. (A few scholars give great weight to newer crops like turnips, but such matters were of very minor importance—setting aside the much earlier introduction of the potato—compared to such things as the increased use of capital and purchased input nutrients. The fact that farmers in western Europe learned how to increase their production while decreasing labor inputs is not at all novel in the history and geography of agriculture. Thus the agricultural revolution of the seventeenth and eighteenth centuries can be considered an effect, not a cause, in the industrializing and urbanizing process.) So the technological side of the Industrial Revolution was not primary cause except as it may have been primary cause for a hemisphere-wide and very slow transformation, as we discussed previously. It appeared too late.

A somewhat similar argument can be given in response to the thesis that the development, by capitalism, of mass wage labor in manufacturing production was primary cause of the Industrial Revolution. This argument is most commonly put forth by those Marxist economists who hold to a strict construction of one of the arguments in Marx's *Capital*. It is indisputable that you cannot have fully mature industrial capitalism without basing it in a wage-labor setting that is also a (relatively) free labor market, one in which workers can go from employer to employer. But these conditions did not exist prior to the late eighteenth century. Wage labor was predominant, but little of it was employed in manufacturing, and hardly ever did a worker confront a really free labor market, with real choices as to place of employment. These were features of industrial capitalism as it emerged after the Industrial Revolution really got rolling.

All such theories about the causes of the Industrial Revolution are diffusionist in the sense, and to the degree, that they see the process as an internal evolution within European history and society. As we noted in Chapter 2, an antidiffusionist, anti-Eurocentric body of historical theory has been developing over the past 50-odd years, a body of theory developed mainly, but not solely, by scholars from the extra-European world. In no other arena of historical discussion has this emerging critical school had as great an impact as it has had on the debates about the origins of the Industrial Revolution.

The thesis that industrial development in Europe depended in many ways on colonial processes was widely accepted in the eighteenth and early nineteenth centuries.[51] Later, perhaps because of the growth of diffusionist ideology with its guiding proposition that Europe is the autonomous source of progress, this thesis fell into disfavor among European historians.[52] It was forcefully advanced by a number of colonial scholars in the 1930s and 1940s. Perhaps understandably it was Indian scholars who emphasized the fact that a highly developed Indian cotton textile industry not only provided some of the new technology for Britain's industry, particularly in dyeing, but also had to be forcibly suppressed by Britain—in a process which some Indian scholars call "the de-industrialization of India"—in order to allow the British industry to develop in the late eighteenth and nineteenth centuries.[53] (The cotton textile industry was the leading sector in the early Industrial Revolution.) Also in the 1930s, West Indian scholars, notably C. L. R. James and Eric Williams, began to advance the thesis that slave-based industry and the slave trade were crucial causal forces in British and French industrialization. This general argument evolved into a broad theory which is now widely argued both by Caribbean scholars—it is sometimes called "the Caribbean school of history," rather too narrowly—and by others, many of whom are African-American and African scholars. This theory is of great importance, and I will try to summarize it briefly, ignoring a number of secondary disagreements among some of its proponents.

The most basic and general argument, advanced first by C. L. R. James and Eric Williams, was the proposition that the West Indian slave-based plantation system in the seventeenth and eighteenth centuries was a highly advanced form of industrial system, implicitly the most advanced form in existence at that time. They, and later writers in the same scholarly tradition, showed that the plantation system involved very heavy capitalization, complex business organization, very advanced industrial technology (in milling, rum manufacture, transport, and so on), a large labor force in the sugar factory as well as in the fields, a considerable force of free workers and supervisors as well as slaves, and, most important of all, immense profits—profits not only from the plantation and its production but also from the slave trade and many ancillary components of what Williams called "the triangular trade."[54] (Said James in his classic history of the Haitian revolution, *The Black Jacobins*, "the slave-trade and slavery were the economic basis of the French revolution . . . Nearly all the industries which developed in France during the eighteenth century had their origin in goods or commodities destined either for the coast of Guinea or for the Americas."[55]) I would extend this argument to a slightly more general proposition: Within the overall economic space which the Europeans controlled in the seventeenth and eighteenth centuries, they found it possible to advance the capitalist industrial production system—large-scale, organized, semimechanized—to its highest level, for that era, *mainly* in the plantation system, using slave labor, until the evolution of industrial production as an overall system had evolved sufficiently so that profits could be made even when the labor force was paid a living wage, a wage permitting subsistence and reproduction of the working class, and the system could then be centrated, imported into Europe itself.[56] Stated differently, the earliest phase of the Industrial Revolution was so crude, undeveloped, and indeed barbaric that free labor could not be used, if the output was to be profitable. Therefore, the capture and forced labor of slaves was necessary for production, or, alternatively, colonial rule elsewhere (as in India) was needed to force the delivery of commodities at very low prices.

Both James and Williams argued that the profits from this complex were crucial in providing much, perhaps most, of the capital required in the early phase of the Industrial Revolution. Williams's book, *Capitalism and Slavery*, provided the classic statement of and argument for this thesis. He showed in great detail how the profits from the slave trade, the slave plantation, and the ancillary economic sectors flowed into England and then into the forms of investment that fueled the Industrial Revolution and its infrastructure (canals, ports, and the like). Most of the mainstream (European) community of scholars has rejected this theoretical position. The general view is that the industrial revolution was an almost entirely intra-European phenomenon, and such matters as the slave trade, the slave plantation, and the profits from all of this *had* to be merely a detail or footnote.[57] Periodically, efforts have been made to refute the theory, but the only part of the theory which has really been subjected to empirical critique is the most limited and in a way least crucial part. Engerman and some others tried to show that, if various assumptions of neoclassical economics are made about the eighteenth-century British economy, and if traditionally low calculations are used as regards the number of slaves brought to America, then it would appear that the slave trade was not really very profitable. But in fact the slave trade itself was only a part of the overall complex that Williams and others were looking at; in fact, the plantation as in industrial system was much closer to the center of their attention because it was here that labor was put to use in generating mass commodities. Inikori and others have shown that the numbers of slave transported to the Americas has been underestimated. Finally, the neoclassical assumptions (among them the argument that there were "normal" profits in an eighteenth-century industry, as though the Industrial Revolution and factor and product markets had already matured) are widely questioned.

Another stream of criticism has come from some Marxists, among them Brenner and Laclau, who share the Eurocentric–diffusionist views of the conventional historical school just discussed.[58] Their positions tend to be grounded in two arguments, one of which is dogmatic and the other fallacious. First, they claim that unfree labor cannot, *by definition*, be considered part of capitalism. This has been answered by C. L. R. James who showed that the error is that of trying to judge a seventeenth- and eighteenth-century labor system by the standards of the mid-nineteenth century, the era of mature competitive capitalism as described by Marx. Even more effective has been the demonstration by Immanuel Wallerstein that capitalism uses a range of alternative labor systems under alternative production conditions, and forced labor is one of these alternative systems.[59] Second, the Marxist critics claim that processes that occurred outside of Europe and involved then the import of commodities and capital into Europe, must be denominated "exchange" rather than "production," and so cannot be considered crucial for industrial development or capitalism. This thesis is simply false: production on a slave plantation is just as much production as is production in a Birmingham needle factory.

Scholars such as Bailey, Beckles, Darity, Mintz, Sheridan, Solow, Robinson, and Rodney, and (on a world-scale canvas) Amin, Wallerstein, and Frank, have, in recent years, given strong backing to the critical theory I have outlined here.[60] Conventional historians sometimes label it "the Williams thesis." My point is that this "thesis" is something much larger: it is the current state of the body of theory that pays adequate attention to the role played by colonialism in the Industrial Revolution.

One other point of contention concerns the significance of demand. All parties concede that the decisions to increase productive capacity, decisions which, in aggregate, led to the Industrial Revolution, were made on the basis of judgments that additional commodities, if produced, could be sold. The conventional historians generally treat the increase in demand as a somehow natural product of the modernization of Europe.[61] The critical historians insist, rather, that colonialism was itself required in order to increase the level of demand such that industrialists would make efforts to increase capacity, efforts which, when the revolution got truly underway, involved the use of powerful new productive technology. The critical historians have indeed shown that an immense amount of demand was generated by the slave trade, by the plantations (demand for food, clothing, machinery, ships, and so on), and by the overall expansion of the trading sphere in which European commodities moved in the eighteenth century and thereafter. I would generalize the case as follows: there would not have been an Industrial Revolution had it not been for the immense demand that Europeans were able to generate in the colonies, and it was this fact that, more than anything else, pushed the Industrial Revolution forward.

Capitalism arose as a world-scale process: as a world system. Capitalism became centrated in Europe because colonialism gave Europeans the power both to develop their own society and to prevent development from occurring elsewhere. It is this dynamic of development and underdevelopment which mainly explains the modern world.

In this chapter and the two preceding ones I have tried to show, with empirical evidence, that there was no "European miracle." Africa, Asia, and Europe shared equally in the rise of capitalism prior to 1492. After that date, Europe took the lead. This happened, as I have tried to demonstrate in this chapter, because of Europe's location near America and because of the immense wealth obtained by Europeans in America and later in Asia and Africa—not because Europeans were brighter or bolder or better than non-Europeans, or more modern, more advanced, more progressive, more rational. These are myths of Eurocentric diffusionism and are best forgotten.

NOTES

1. Europeans did not "discover" America: the hemisphere was settled many millennia earlier by people who migrated in from Siberia and the Arctic. So I prefer not to conceptualize the European arrival as a "discovery." Likewise, the idea that the

Western Hemisphere is a "New World" is false since it was hardly new to those who lived there and greeted Columbus on his arrival in 1492. It is, however, very difficult to avoid using the phrase "New World" in certain contexts, and I will occasionally have to do so.

2. See K. N. Chaudhuri, *Trade and Civilization in the Indian Ocean* (1985); Simkin, *The Traditional Trade of Asia* (1968); Sherif, *Slaves, Spices and Ivory in Zanzibar* (1987). It is highly likely that West Africans sailed across to the Americas before 1492. (See DeVisse and Labib, "Africa in Intercontinental Relations," 1984.) However, because there seem not to have been major mercantile–maritime port cities in West Africa—unlike East Africa—it is not likely that transatlantic voyages before 1492 had significant impact on Africa or on America. This is probable for several reasons. First, in the absence of such major port cities and large-volume long-distance sea commerce, it is likely that ships along the coast were rather small. They would easily have been capable of a westward voyage to America, given the strong and steady trade winds blowing westward, but a return voyage would have to have been made far to the north or the south, in the zone of the westerlies—roughly as far north as the latitude of southern Europe or as far south as the latitude of Namibia. Therefore the round trip would have been a very formidable undertaking. On the other hand, we may indeed learn from future scholarship that West African sailors, Moroccan sailors, and West European sailors all were fishing and whaling quite regularly off the coast of America (perhaps on the Grand Banks) before 1492; if strong evidence for this emerges, then we would consider it likely that West Africans were familiar with the round-trip voyage and with some parts of the American coast. But we do not have such evidence at present, and we have to consider it more likely that any West African ships that reached America were blown off course, in which case a return voyage would have been very difficult. It would have to have been made without prior knowledge of the long, circuitous route (unless American sailors knew the route and gave navigational information to the African sailors—but we do not have convincing evidence at present that Americans crossed the Atlantic before 1492). Secondly, the portion of the American coast that is closest to Africa, roughly the Brazilian coast south of the mouth of the Amazon, seems not to have had major population concentrations and abundant gold and silver artifacts inviting trade or plunder. (Granted, if West Africans reached the West Indies, they would have found such artifacts in abundance, as did Columbus.) And thirdly, the complex of historical conditions that would turn a single voyage into the beginning of a massive conquest seem unlikely to have been present in coastal West Africa. Large-scale trade, a class of merchant–capitalists, banking and other institutions of capitalism, and the like, were found in interior West African urban centers, but not, it seems, in the urban centers along the coast: these were not important as mercantile–maritime centers. For the interior cities, major long-distance trade went northward and eastward overland, and it is unlikely that attempts would have been made to develop large-scale oceanic travel from a coastal harbor.

Some scholars maintain the truth of two propositions about West African transatlantic voyaging that I cannot accept. The first of these asserts that Africans exerted an important influence on American cultures before 1492. The second asserts that West Africans crossed the Atlantic in much the same way as did Columbus, but they had different values than the Europeans, and did not choose to murder, plunder, enslave, and enrich themselves at the expense of Americans as did the Europeans; and therefore they did not attempt conquest. Most of the evidence offered in support of important precolumbian diffusion from Africa to America is taken from the old arguments of European scholars of the "extreme diffusionism" school we discussed in Chapter 1. The extreme diffusionists claimed that ancient Egyptians or Phoenicians crossed the Atlantic, and, in essence, brought civilization to the Americas. Some modern scholars modify this mainly by insisting that because Egypt was a clearly African civilization—this I am certain is true—it was an African people, not a putatively European people, who brought civilization to the Americas. A second source of evidence is the apparently African facial features of the great Olmec head sculptures of southern Mexico. But some precolumbian Americans must have had these features, too: they are not rare among modern Latin American Indians. But the most serious objection to this theory is the following: The Olmec civilization is the oldest known civilization in the Americas. If Olmec civilization came from Africa, and was not developed indigenously by people of America, then we would have to say that Americans simply did not have the ability to civilize themselves; civilization had to be something brought in from elsewhere, by diffusion. This is viewed as a deep insult by Latin Americans, who maintain, I am sure correctly, that Western Hemisphere peoples developed civilization on their own. Perhaps they acquired a few domesticates from sailors arriving from across the Atlantic or the Pacific. But the real cultural development was a matter of independent invention, not diffusion. Again we notice that the form of argument comes from classical diffusionism: some human communities are inventive and others merely imitative. Based only on the thin and questionable evidence that has been presented thus far that Africans brought major cultural advances to America, this thesis is not persuasive.

Somewhat more troublesome to me is the argument that when Africans crossed the Atlantic before or at the same time as Columbus did, they did not have the savage values of the Europeans, and so did not try to conquer, loot, and enslave. To accept this, one would have to believe that there is something absolutely fundamental in European culture, something very old, and very deeply embedded, that makes Europeans different from other humans. This admits a good part of the Eurocentric claim that Europeans are unique among humans; it merely inverts the argument and claims that their uniqueness lies not in progressiveness but in aggressiveness, predatoriness, and cupidity. I am much more comfortable with an argument that begins with the idea of a common basal human mentality ("psychic unity"). It then explains the bloodthirstiness of the European conquistadors as an effect of the kind of civilization they represented: its development of an oppressive class structure in feudalism, and its further development of protocapitalism, a system in which wealth is obtained at all cost and in any way possible. Bloodthirsty protocapitalist communities, ready and anxious to conquer, loot, and enslave wherever this brought a profit, were found in many parts of the Eastern Hemisphere, in all three continents. My argument in this book is that the key factor favoring European moves of conquest in the Americas, and not favoring West African moves of this sort, was the existence of major mercantile-*maritime* centers in coastal Europe, protocapitalist centers of the sort found in the interior of Africa but with the *added* features associated with long-distance *oceanic* trade. Sofala and Kilwa in East Africa had these features but Sofala and Kilwa were—as we note in this chapter—very much farther from the American looting grounds than were the Iberian ports and the Canary outports. (I have not cited specific scholars who hold these views that I criticize because a full and fair review of their theories is not possible in the space of a single long footnote. Obviously, I do not agree with the theory of Ivan Van Sertima, as presented in his important work *They Came Before Columbus* [1976], concerning the precolumbian diffusion of major civilizing traits from Africa to America, although he is very probably correct in his view that Africans *did* come to America before Columbus did.)

3. Filesi, *China and Africa in the Middle Ages* (1972); Ma Huan, *The Overall Survey of the Ocean's Shores* (1970); Panikkar, *Asia and Western Influence* (1959).

4. I have not learned of documented evidence that North Africans or West Africans regularly sailed up and down the coast past Cape Bojador. (See DeVisse and Labib, "Africa in Intercontinental Relations," 1984.) Apparently medieval sailing techniques—European and non-European—had difficulty with the passage prior to the time when Portuguese voyages began in the fifteenth century. However, there was no question of "discovery." The sea route was known in antiquity. Important land routes paralleled the entire length of the coast, from Fez south to Takrur (near modern Dakar) and beyond (see Niane, "Mali and the Second Mandingo Expansion," 1984, and Levitzion, "The Early States of the Western Sudan to 1500," 1971), and there were medieval-era settlements in the Canaries and along the coast itself. Basically, it was much cheaper to travel overland, and probably faster. What the Portuguese "discovered" was a method of outflanking the competing merchant interests of North and West Africa, by applying sailing technology known to Europeans and East Africans but not known (or at any rate used, or at any rate *known* to have been used) by West Africans in that period. It should be noted also that the Portuguese navigational strategy in passing Bojador was basically the same strategy used for voyages to the Atlantic islands, and probably known to Moroccan sailors as well as Europeans.

5. Blaut, "Diffusionism: A Uniformitarian Critique" (1987).

6. For general reviews, see Crosby, *The Columbian Exchange* (1972) and McNeill, *Plagues and Peoples* (1976). See Borah and Cooke, "La Demografía Histórica de América Latina: Necesidades y Perspectivas" (1972); Whitmore, "A Simulation of Sixteenth-Century Population Collapse in the Basin of Mexico" (1991); Alchon, *Native Society and Disease in Colonial Ecuador* (1991); Lovell, " 'Heavy Shadows and Black Night': Disease and Depopulation in Colonical America" (1992).

7. See Denevan, *The Native Population of the Americas in 1492* (1976), for a review of the disputes concerning the American population at the time of the conquest.

8. Crosby, *The Columbian Exchange* (1972); Alchon, *Native Society and Disease in Colonial Ecuador* (1991).

9. For a discussion of the various calculations, see Denevan, *The Native Population of the Americas in 1492* (1976); Denevan, "The Pristine Myth: The Landscape of the Americas in 1492" (1992); Lovell, " 'Heavy Shadows and Black Night' " (1992); and Whitmore, "A Simulation of Sixteenth-Century Population Collapse in the Basin of Mexico" (1991).

10. The assumption here is that population continued to grow so long as food resources for hunting, gathering, fishing, and shellfishing were abundant. At a certain purely hypothetical time, it is likely that people who had certainly already experimented with crop cultivation, found that a better supply of food (and fiber, etc.) would be obtained through agriculture, and so began the transformation. Note that this argument is in no way Malthusian.

11. See Fiedel, *Prehistory of the Americas* (1987).

12. See Crosby, *The Columbian Exchange* (1972), Lovell, " 'Heavy Shadows and Black Night,' " and Alchon, *Native Society and Disease in Colonial Ecuador* (1991).

13. Miskimin, *The Economy of Early Renaissance Europe, 1300–1460* (1969).

14. Abel, *Agricultural Fluctuations in Europe from the Thirteenth to the Twentieth Centuries* (1980).

15. de Vries, *European Urbanization, 1500–1800* (1984).

16. Hodgett, *A Social and Economic History of Medieval Europe* (1972) ("[The] 200 years after c.1320 may be said to be a period of down-turn in the [European] economy as a whole," p. 212); Lopez and Miskimin, "The Economic Depression of the Renaissance," (1961–1962); C. T. Smith, *An Historical Geography of Western Europe Before 1800* (1969).

17. Lopez, "Hard Times and Investment in Culture" (1953); Thorndyke, "Renaissance or Prenaissance?" (1943).

18. Braudel, "Prices in Europe from 1450 to 1750" (1967); de Vries, *European Urbanization* (1984). On the rapid impact of these changes on Asia, see, for example, Atwell, "International Bullion Flows and the Chinese Economy *circa* 1530–1650" (1982); Aziza Hasan, "The Silver Currency Output of the Mughal Empire and Prices in India during the 16th and 17th centuries" (1969).

19. Céspedes, *Latin America: The Early Years* ((1974); McAlister, *Spain and Portugal in the New World, 1492–1700* (1984).

20. In classical political economy, and in some of modern Marxist economics, the idea of "primitive accumulation" serves as a kind of catchall for ways of accumulating capital that did not involve—in essence—capitalist enterprise. Pirate treasures and the like were "primitive accumulation," and, in general, the kind of wealth brought from the American colonies in the sixteenth century was considered primitive accumulation. (According to Marx, "The treasures captured outside Europe by undisguised looting, enslavement and murder flowed back to the mother-country and were turned into capital there": Marx, *Capital*, 1976, vol. 1, p. 918). But "primitive accumulation" cannot really be defined with any precision. I will argue here that the wealth accumulated in the Americas was primitive only in the sense that it was part of a preindustrial-capitalist economy. In other respects, notably in the involvement of labor and value produced by labor, it was regular accumulation. The distinction is very fundamental, as we will see, because, if what transpired in the colonies was not "real" or "ordinary" accumulation, scholars can claim that the colonial economy was backward and "feudal," rather than a primitive sort of capitalism.

21. E. J. Hamilton, *American Treasure and the Price Revolution in Spain, 1501–1650* (1934); Brading and Cross, "Colonial Silver Mining: Mexico and Peru" (1972); H. and P. Chaunu, *Séville et l'Atlantique (1504–1650)*, vol. 6, pt. 1 (1956); Cross, "American Bullion Production and Export 1550–1750" (1983).

22. See note 19 above.

23. Vicens Vives, *An Economic History of Spain* (1969).

24. Vilar, *A History of Gold and Money, 1450–1920* (1976).

25. There remained, nonetheless, a very large trade carried on by non-European merchants in the China Seas and the Indian Ocean.

26. E. J. Hamilton, "American Treasure and the Rise of Capitalism" (1929), and *American Treasure and the Price Revolution in Spain* (1934). Also see the important book by Walter Prescott Webb, *The Great Frontier* (1951), which builds in part on Hamilton's theory to argue for the great importance of the Americas in the rise of Europe during this period and later.

27. Galloway, *The Sugar Cane Industry: An Historical Geography from its Origins to 1914* (1989); Deerr, *The History of Sugar* (1949–1950).

28. See, for example, Galloway, *The Sugar Cane Industry* (1989); Deerr, *The History of Sugar* (1949–1950); Watson, *Agricultural Innovation in the Early Islamic World: The Diffusion of Crops and Farming Techniques, 700–1100* (1983); N. S. Gupta,

Industrial Structure of India During the Medieval Period (1970); Niane, ed., *UNESCO General History of Africa*, Vol. 4. (1984); Bray, *Science and Civilization in China, Vol. 6, Part 2, Agriculture* (1984).

29. Deerr, *The History of Sugar* (1949–1950).

30. Simonsen, *História Econômica do Brasil, 1500–1820* (1944); Furtado, *The Economic Growth of Brazil* (1963); Minchinton, *The Growth of English Overseas Trade* (1969). Also see the more general, but extremely important, works of I. Wallerstein, A. G. Frank, and S. Amin, particularly Wallerstein's *The Modern World System*, 3 vols. (1974–1988), Frank's *Capitalism and underdevelopment in Latin America* (1968) and his *World Accumulation, 1492–1789* (1978), and Amin's *Accumulation on a World Scale* (1974) as well as his *Unequal Development* (1976).

31. Edel, "The Brazilian Sugar Cycle of the 17th Century and the Rise of West Indian Competition"(1969).

32. Furtado, *Economic Growth of Brazil* (1963).

33. See K. N. Chaudhuri, *Trade and civilization in the Indian Ocean* (1985); Satish Chandra, *The Indian Ocean: Explorations in History, Commerce and Politics* (1987); Magalhães-Godinho, *L'Economie de L'Empire Portugais aux XV et XVI Siècles* (1969).

34. de Vries, *European Urbanization* (1984).

35. William Denevan, "Introduction," in Denevan, ed., *The Native Population of the Americas in 1492* (1976), and "The Pristine Myth: The Landscape of the Americas in 1492" (1992; Lovell, " 'Heavy Shadows and Black Night' " (1992).

36. See Borah and Cook, "La demografía histórica de América Latina: necesidades y perspectivas" (1972); Whitmore, "A Simulation of Sixteenth-Century Population Collapse in the Basin of Mexico" (1991).

37. Semo, *Historia del Capitalismo en México: Los Orígenes, 1521–1763* (1982).

38. Radell, "The Indian Slave Trade and Population of Nicaragua During the Sixteenth Century" (1976).

39. Bakewell, "Mining in Colonial Spanish America" (1984).

40. Semo, *Historia del Capitalismo* (1982).

41. For various calculations, see Curtin, *The Atlantic Slave Trade* (1969); Furtado, *Economic Growth of Brazil* (1963); Deerr, *History of Sugar* (1949–1950); Florescano, "The Formation and Economic Structure of the Hacienda in New Spain"; Inikori, *The African Slave Trade from the Fifteenth to the Nineteenth Century* (1979), esp. pp. 57 and 248; McAlister, *Spain and Portugal in the New World (1984)*.

42. McAlister, *Spain and Portugal in the New World* (1984).

43. In the present discussion I am giving far too little attention to Africa and particularly to the effects of the slave trade in Africa. See Chapter 2.

44. Fisher, "The Price Revolution: A Monetary Interpretation" (1989).

45. Dunn, *Sugar and Slaves: The Rise of the Planter Class in the English West Indies, 1624–1713* (1972), pp., 10–11.

46. Tawney, *Religion and the Rise of Capitalism* (1952 edition).

47. Hannemann, *The Diffusion of the Reformation in Southwestern Germany, 1518–1534* (1975).

48. There is debate as to why the center shifted from Iberia to the lower Rhine–southern England region. Perhaps the same forces which had made this northern region a mercantile–maritime center in the Middle Ages permitted it to gain control of the overseas enterprise: namely, large population, abundant nearby fertile land and forest resources, access to many markets (the Rhine, the Baltic, etc.). Vis-à-vis Italy, it held most of these same advantages plus that of location on the Atlantic and possession of the requirements for rapid growth of oceanic shipping and fishing fleets.

49. See Deerr, *The History of Sugar* (1949–1950); Curtin, *The Atlantic Slave Trade* (1969); and Inikori, *The African Slave Trade from the Fifteenth to the Nineteenth Century* (1979). The question whether slave labor is or is not proletarian—a serious issue in discussions of the slave plantation system (see Mintz, *Sweetness and Power: The Place of Sugar in Modern History*, 1985)—will be taken up later in this chapter. In any event there is no disagreement about the contribution of slave (and other forced) labor to capital accumulation, hence to surplus value generation, using "surplus value" in a sense appropriate to modes of production different from industrial capitalism.

50. However, a good share of the old landowning elite joined in the new enterprise. It is not correct to assume that the new protocapitalist elite was in simple opposition to the old elite. There is much confusion on this matter, some of it occasioned by literal acceptance of Marx's idea that merchants are somehow not the class that evolves into the early capitalist, entrepreneurial, accumulating class. On the role of medieval merchants in protocapitalism, see Thrupp, *The Merchant Class of Medieval London (1300–1500)* (1948); Carus-Wilson, *Medieval Merchant Venturers* (1967).

51. See R. W. Bailey, "Africa, the Slave Trade, and the Rise of Industrial Capitalism in Europe and the United States: A Historiographic Review" (1986); W. Darity, Jr., "British Industry and the West Indies Plantations" (1990).

52. There were, of course, exceptions. Brooks Adams, in his 1895 work *The Law of Civilization and Decay*, argued (pp. 259–260) that the British victory at Plassey in 1757, which immediately gave Britain access to cheap Indian cotton (and other Indian "plunder") set into motion the explosive industrialization of Britain's cotton textile industry, leading directly and immediately to the major inventions of that industry: the spinning jenny in 1764, the mule in 1776, and Watt's steam engine in 1768.

53. See Palme Dutt, *The Problem of India* (1943); Alavi et al., *Capitalism and Colonial Production* (1982).

54. C. L. R. James, *The Black Jacobins: Toussaint L'Ouverture and the San Domingo Revolution* (1938) and *A History of Negro Revolt* (1938); Eric Williams, *Capitalism and Slavery* (1944). Also see the later work by James, "The Atlantic Slave Trade and Slavery: Some Interpretations of their Significance in the Development of the United States and the Western World" (1970); and the later work by Williams, *British Historians and the West Indies* (1966). Important recent contributions include: R. W. Bailey, "The Slave(ry) Trade and the Development of Capitalism in the United States: The Textile Industry of New England" (1990); W. Darity, "British Industry and the West Indian Plantations" (1990); J. Inikori, "Slavery and the Revolution in Cotton Textile Production in England " (1989). Also see note 60 below.

55. *The Black Jacobins* (1938), pp. 47–48.

56. I have argued elsewhere (Blaut, *The National Question*, 1987b, chap. 7) that the level of oppression and exploitation associated with slave labor as it was used in the plantations could not have been applied to members of the European cultural community itself. (This was indeed tried, but quickly abandoned in favor of slave labor.) Generally, cultural rules and practices limit the level of exploitation of producers within a society—a matter of maintaining social peace in a society—but no such rules apply to external or foreign workers.

57. See the excellent review by C. Robinson, "Capitalism, Slavery and Bourgeois Historiography" (1987). Also excellent is Bailey, "The Slave(ry) Trade and the Development of Capitalism in the United States" (1990).

58. See Brenner's "The Origins of Capitalist Development: A Critique of Neo-Smithian Marxism" (1977); E. Laclau, *Politics and Ideology in Marxist Theory* (1977).

59. Wallerstein, *The Modern World System* (1974–1988).

60. See Bailey, "The Slave(ry) Trade and the Development of Capitalism in the United States" (1990), Darity, "British Industry and the West Indian Plantations" (1990), Mintz, *Sweetness and Power: The Place of Sugar in Modern History* (1985), and Robinson, "Capitalism, Slavery and Bourgeois Historiography" (1987); also, see Beckles, "'The Williams Effect': Eric Williams' *Capitalism and Slavery* and the Growth of West Indian Political Economy" (1987); Sheridan, *Sugar and Slavery* (1973), and his "Eric Williams and *Capitalism and Slavery*: A Biographical and Historiographical Essay" (1987); Solow, "Capitalism and Slavery in the Exceedingly Long Run" (1987); Inikori, "Slavery and the Development of Industrial Capitalism" (1989); Rodney, *How Europe Underdeveloped Africa* (1972).

61. Some Marxists treat it this way, too: "[What] distinguished the English industrial development of the early modern period was its continuous character, its ability to sustain itself and to provide its own self-perpetuating dynamic. Here . . . the key was to be found in the capitalist structure of [English] agriculture" (Brenner, "Agrarian Class Structure and Economic Development in Pre-Industrial Europe," 1985, p. 53).

CHAPTER 5

Conclusion

This book has two basic themes or arguments. First, in Chapter 1, I try to explain what Eurocentric diffusionism is as a body of ideas, and to show how this theory—or supertheory, or world model—came to dominate European scholarly thought a century ago and why it still does so to a considerable extent today. And second, in Chapters 2 through 4, I carefully examine the single most important part of diffusionism, the theory of Europe's historical superiority or priority, the theory of "the European miracle," and I try to refute it.

Diffusionism needs to be analyzed much more thoroughly than I have been able to do in this book. Many diffusionist theories and programs that, today, exert an important and unfortunate influence on many fields of thought and action have not been discussed here. In other writings I have explored the influence of diffusionism on theory and practice concerning the national question, or nationalism,[1] and on theory and practice concerning the development of peasant agriculture.[2] Other writers have, of course, examined many aspects of diffusionism and problems caused by diffusionism.[3] But, overall, the critique of diffusionism has barely begun.

The critique will have to range across many fields of scholarship and practice. Here—just to make this point clear—are four examples.

1. Philosophical dualism, the body of epistemological and ontological doctrine developed in European thought from Descartes to Kant and the neo-Kantians, appears to be, in part, a projection of the dualism of Inside and Outside. Reason is Inside. Mere matter, mere sensuousness, is Outside—the non-European world and the irrational mentation of its inhabitants.

2. The so-called Big Bang Theory, the theory that everything began at one space–time point and this point was *here*, seems to be diffusionism

on the largest canvas of all. Big Bang cosmogony appears to be fortified less by empirical evidence than by a hunch that the whole idea is "reasonable"—the essential judgment (as we noticed in Chapter 1) by which culture projects its prejudices into science.[4]

3. The theory that AIDS diffused out of Africa is very reminiscent of a historical chain of theories, each explaining some plague as a counterdiffusion from non-Europe to Europe. (We discussed aspects of this question in Chapters 1 and 2.) A recent book entitled *AIDS, Africa and Racism* gives important evidence that the AIDS-out-of-Africa doctrine may, indeed, be simply a new incarnation of this diffusionist view of human disease.[5] If this is the case, the matter of causality of HIV-retrovirus-caused disease may have to be rethought. The forms found *outside* of Africa may be more relevant for explanation and cure than those *inside* Africa.

4. Many theories about economic history since the beginning of the Industrial Revolution, and about economic development today, seem to be steeped in diffusionism. The Industrial Revolution has *not* diffused outward from Europe to non-Europe. Not only does it have origins in non-Europe as well as Europe (we discussed this matter in Chapter 4), but the notion that industrialization has been spreading to the non-European world is largely a false (conformal) idea. The diffusion of *maquiladora*-style assembly-plant activities in the Third World is not genuine industrialization but rather a kind of world-scale putting-out system: Outside provides cheap labor, Inside provides most of the raw materials and most of the consumption, and garners nearly all of the profit as well as the permanent infrastructure. The industrialization of Japan began long ago and was not an effect of diffusion.[6] The industrialization of Korea and one or two East Asian ministates in recent decades has not been imitated elsewhere.[7] The diffusion of industrialization, therefore, is not a simple and natural diffusion process, but a political agenda. And an agenda for scholarly inquiry.

This book, therefore, has no real conclusion. The book itself is an introduction: an introduction to the study, to the diagnosis and treatment, of a serious malady of the mind.

NOTES

1. Blaut, *The National Question* (1987b); Blaut and Figueroa, *Aspectos de la cuestión nacional en Puerto Rico* (1988).
2. Blaut, "Two Views of Diffusion" (1977) and "Diffusionism: A Uniformitarian Critique" (1987a).

3. Reference has been made to this work in Chapters 2, 3, and 4.

4. See Talkington, "But the Editor Looks at the Universe from a Different Frame of Reference" (1986); Frankel, "Marxism and Physics: A New Look" (1991).

5. Chirimuuta and Chirimuuta, *AIDS, Africa and Racism* (1989). A naively diffusionist view of AIDS is given in Shannon and Pyle, "The Origin and Diffusion of AIDS" (1989); see the critique of this view in Watts, Okello, and Watts, "Medical Geography and AIDS" (1990).

6. Japan became industrialized precisely because of a lack of diffusion. It was the only major non-European country that managed to avoid European domination, and this resulted from its inaccessibility. It was, among major societies, the farthest and most inaccessible from the standpoint of Europeans, and by the time European power had subdued China, in the nineteenth century, Japan had been able to begin its military modernization; hence the victory over Russia, the beginnings of colonial expansion, and the onset of an industrial revolution around 1900.

7. At the other end of the scale, giant countries like India and Brazil have a great deal of industry but in proportion to their size—and on a per capita measurement—they are no more industrialized than are smaller Third world countries. See Amin, *Delinking: Toward a Polycentric World* (1990).

Bibliography

Abel, W. (1980). *Agricultural fluctuations in Europe from the thirteenth to the twentieth centuries*. London: Methuen.

Abu-Lughod, J. (1987–1988). The shape of the world system in the thirteenth century. *Studies in Comparative International Development* 22(4):3–25.

———. (1989). *Before European hegemony: The world system* A.D. *1250–1350*. New York: Oxford University Press.

Adams, B. (1895). *The law of civilization and decay*. New York: Macmillan.

Adams, W., Van Gerven, D., and Levy, R. (1978). The retreat from migrationism. *Annual Review of Anthropology* 7:483–532.

Ahn, P. (1970). *West African soils*. London: Oxford University Press.

Ajayi, J., and Crowder, M., eds. (1972). *History of West Africa*. Vol. 1. New York: Columbia University Press.

Alavi, H. (1982). India: The transition to colonial capitalism. In H. Alavi, et al., eds., *Capitalism and colonial production*. London: Croom Helm.

———, Burns, P., Knight, G., Mayer, P., and McEachern, D., eds. (1982). *Capitalism and colonial production*. London: Croom Helm.

Alchon, S. (1991). *Native society and disease in colonial Ecuador*. Cambridge, England: Cambridge University Press.

Amin, S. (1973). *Neo-colonialism in West Africa*. New York: Monthly Review Press.

———. (1974). *Accumulation on a world scale*. New York: Monthly Review Press.

———. (1976). *Unequal development*. New York: Monthly Review Press.

———. (1985). Modes of production: History and unequal development. *Science and Society* 49:194–207.

———. (1988). *Eurocentrism*. New York: Monthly Review Press.

———. (1990). Colonialism and the rise of capitalism: A comment. *Science and Society* 54:67–72.

———. (1990). *Delinking: Toward a polycentric world*. London: Zed Books.

———. (1992). On Jim Blaut's "Fourteen ninety-two." *Political Geography* 11:394–396.

Anderson, J. (1882). *New manual of general history*. New York: Clark and Maynard.

Anderson, P. (1974). *Passages from antiquity to feudalism*. London: New Left Books.

———. (1974). *Lineages of the absolute state*. London: New Left Books.

Appadorai, A. (1936). *Economic conditions in southern India (1000–1500* A.D.*)*. 2 vols. Madras: University of Madras Press.

Asad, T., ed. (1975). *Anthropology and the colonial encounter*. London: Ithaca Press.

Aston, T., and Philpin, C., eds. (1985). *The Brenner debate: Agrarian class structure and economic development in pre-industrial Europe.* Cambridge, England: Cambridge University Press.

Atwell, W. (1982). International bullion flows and the Chinese economy *circa* 1530–1650. *Past and Present* 95:68–91.

Baechler, J. (1988). The origins of modernity: Caste and feudality (India, Europe and Japan). In J. Baechler, J. A. Hall, and M. Mann, eds. *Europe and the rise of capitalism.* Oxford: Basil Blackwell.

———, Hall, J. A., and Mann, M., eds. (1988). *Europe and the rise of capitalism.* Oxford: Basil Blackwell.

Bailey, A., and Llobera, J., eds. (1981). *The Asiatic mode of production: Science and politics.* London: Routledge and Kegan Paul.

Bailey, R. (1986). Africa, the slave trade, and the rise of industrial capitalism in Europe and the United States: A historiographic review. *American History: A Bibliographic Review* 2:1–91.

———. (1990). The slave(ry) trade and the development of capitalism in the United States: The textile industry of New England. *Social Science History* 14(3):373–414.

Bakewell, Peter. (1984). Mining in colonial Spanish America. In L. Bethell, ed., *The Cambridge history of Latin America: Colonial Latin America,* Vol. 2. Cambridge, England: Cambridge University Press.

Baran, P., and Sweezy, P. (1966). *Monopoly capital.* New York: Monthly Review Press.

Barnes, H. E., ed. (1925). *Ploetz' manual of universal history.* New York: Blue Ribbon Books.

Beckles, H. (1987). "The Williams effect": Eric Williams' *Capitalism and slavery* and the growth of West Indian political economy. In B. Solow, and S. Engerman, eds., *British capitalism and Caribbean slavery: The legacy of Eric Williams.* Cambridge, England: Cambridge University Press.

Berkner, L. (1975). The use and misuse of census data for the historical analysis of family structures. *Journal of Interdisciplinary History* 5:721–738.

———. (1989). The stem family and the developmental cycle of the peasant household, *American Historical Review* 77:398–428.

Bernal, M. (1987). *Black Athena: The Afroasiatic roots of classical civilization,* Vol. 1., *The fabrication of ancient Greece.* London: Free Association Books.

———. (1991). *Black Athena: The Afroasiatic roots of classical civilization,* Vol. 2, *The archeological and documentary evidence.* London: Free Association Books.

Bhatia, B. (1967). *Famines in India.* Bombay: Asia Publishing House.

Bjørklund, O., Hølmhoe, H., Røhr, A., and Lie, B. (1970). *Historical atlas of the world.* New York: Barnes and Noble.

Black, C. (1966). *The dynamics of modernization: A study in comparative history.* New York: Harper and Row.

Blaikie, P. (1978). The theory of the spatial diffusion of innovations: a spacious cul-de-sac. *Progress in Human Geography* 2:268–295.

Blaut, J. (1962). The nature and effects of shifting agriculture. In *Symposium on the impact of man on humid-tropics vegetation.* Canberra: UNESCO and Australian Government Printer.

———. (1963). The ecology of tropical farming systems. *Revista Geográfica* 28:47–67.

———. (1970). Geographic models of imperialism. *Antipode* 2(1):65–85.

———. (1973). The theory of development. *Antipode* 5(2):22–26.

———. (1976). Where was capitalism born? *Antipode* 8(2):1–11.

————. (1977). Two views of diffusion. *Annals of the Association of American Geographers* 67:343–349.

————. (1979). Some principles of ethnogeography. In S. Gale, and G. Olsson, eds., *Philosophy in geography*. Dordrecht: Reidel.

————. (1982). Nationalism as an autonomous force. *Science and Society* 46:1–23.

————. (1984). Modesty and the movement. In T. Saarinen, et al., eds., *Environmental perception and behavior: An inventory and prospect*. Chicago: University of Chicago Department of Geography, Research Paper No. 209.

————. (1987a). Diffusionism: A uniformitarian critique. *Annals of the Association of American Geographers* 77:30–47.

————. (1987b). *The national question: Decolonizing the theory of nationalism*. London: Zed Books.

————. (1989). Colonialism and the rise of capitalism. *Science and Society* 53:260–296.

————. (1989). Review of J. Baechler, J. Hall, and M. Mann, eds., *Europe and the rise of capitalism*. *Progress in Human Geography* 13(3):441–448.

————. (1991). Natural mapping. *Transactions of the Institute of British Geographers* 16(n.s.):55–74.

————. (1991). Review of P. Curtin, *The plantation complex*. *Journal of Historical Geography* 17:472–474.

————. (1992a). The theory of cultural racism. *Antipode* 24(4):289–299.

————. (1992b). Fourteen ninety-two. *Political Geography* 11(3):355–385.

————, and Figueroa, L. (1988). *Aspectos de la cuestión nacional en Puerto Rico*. San Juan: Editorial Claridad.

————, Frank, A. G., Amin, S., Dodgshon, R., Palan, R., and Taylor. P. (1992). *Fourteen ninety-two: The debate on colonialism, Eurocentrism, and history*. Trenton, NJ: Africa World Press.

————, and Ríos-Bustamante, A. (1984). Commentary on Nostrand's "Hispanos" and their "Homeland." *Annals of the Association of American Geographers* 74:157–164.

Blum, J. (1978). *Pseudoscience and mental ability*. New York: Monthly Review Press.

Boas, F. (1938). *The mind of primitive man*. New York: Macmillan.

————. (1948). *Race, language and culture*. New York: Macmillan.

Bois G. (1985). Against the neo-Malthusian orthodoxy. In Aston, T., and Philpin, C., eds., *The Brenner debate: Agrarian class structure and economic development in pre-industrial Europe*. Cambridge, England: Cambridge University Press.

Borah, W., and Cook, S. F. (1972). La demografía histórica de América Latina: necesidades y perspectivas. In J. Bezant et al., eds., *La historia económica en América Latina*. México, D.F.: Sep-Setentas.

Bowler, P. (1989). *The invention of progress*. Oxford: Basil Blackwell.

Brading, D. A., and Cross, H. C. (1972). Colonial silver mining: Mexico and Peru. *Hispanic-American Historical Review* 52:545–79.

Brantlinger, P. (1988). *Rule of darkness: British literature and imperialism, 1830–1914*. Ithaca: Cornell University Press.

Braudel, F. (1967). Prices in Europe from 1450 to 1750. In E. Rich, and C. Wilson, eds. *The Cambridge economic history of Europe, Vol. 4: The economy of expanding Europe in the sixteenth and seventeenth centuries*. Cambridge, England: Cambridge University Press.

————. (1972). *The Mediterranean*. New York: Harper and Row.

Bray, Francesca. (1984). *Science and civilization in China, Vol. 6, part 2, Agriculture.* (Joseph Needham, principal author and editor.) Cambridge, England: Cambridge University Press.

Brenner, R. (1977). The origins of capitalist development: A critique of neo-Smithian Marxism. *New Left Review* 104:25–93.

———. (1985). Agrarian class structure and economic development in pre-industrial Europe. In T. Aston and C. Philpin, eds., *The Brenner debate: Agrarian class structure and economic development in pre-industrial Europe.* Cambridge, England: Cambridge University Press. (Originally in *Past and Present* No. 70, 1976).

———. (1985). The Agrarian roots of European capitalism. In T. Aston, and C. Philpin, eds., *The Brenner debate: Agrarian class structure and economic development in pre-industrial Europe.* Cambridge, England: Cambridge University Press. (Originally in *Past and Present*, no. 97, 1982).

———. (1986). The social basis of economic development. In J. Roemer, ed., *Analytical Marxism.* Cambridge, England: Cambridge University Press.

Browett, J. (1980). Development, the diffusionist paradigm and geography. *Progress in Human Geography* 4:56–79.

Brown, L. (1981). *Innovation diffusion: A new perspective.* London: Methuen.

Buckle, H. T. (1913). *History of civilization in England.* 4 vols., 2d ed., rev. New York: Hearst's International Library.

Bury, J. B. (1932). *The idea of progress.* New York: Macmillan.

Cabral, A. (1979). *Unity and struggle: Speeches and writings of Amilcar Cabral.* New York: Monthly Review Press.

Carter, G. (1968). *Man and the land: A cultural geography.* New York: Holt, Rinehart, and Winston.

Carus-Wilson, E. M. (1967). *Medieval merchant venturers.* 2d ed. London: Methuen.

Césaire, A. (1972). *Discourse on colonialism.* New York: Monthly Review Press.

Céspedes, G. (1974). *Latin America: The early years.* New York: Alfred A. Knopf.

Chambers, M., et al. (1987). *The Western experience.* 2 vols., 4th ed. New York: Alfred A. Knopf.

Chan-Cheung, J. (1967). The smuggling trade between China and Southeast Asia during the Ming Dynasty. In F. Drake, ed., *Symposium on historical, archeological, and linguistic studies on southern China.* Hong Kong: Hong Kong University Press.

Chandra, B. (1981). Karl Marx, his theories of Asian societies, and colonial rule. *Review* 5(1):13–94.

Chandra, S. (1964). Commerce and industry in the medieval period. In B. Ganguli, ed., *Readings in Indian economic history.* Bombay.

———, ed. (1987). *The Indian Ocean: Explorations in history, commerce and politics.* New Delhi: Sage.

Chaudhuri, K. N. (1978). *The trading world of Asia and the English East India Company, 1660–1760.* Cambridge, England: Cambridge University Press.

———. (1985). *Trade and civilization in the Indian Ocean.* Cambridge, England: Cambridge University Press.

Chaudhuri, S. (1974). Textile trade and industry in Bengal Suba, 1650–1720. *Indian Historical Review* 1:262–278.

Chaunu, H., and Chaunu, P. (1956). *Séville et l'Atlantique (1504–1650).* Part 1. Paris: SEVPEN.

Chicherov, A. (1976). On the multiplicity of socio-economic structures in India in the 17th and 18th century. In G. Abramov, ed., *New Indian studies by Soviet scholars*. Moscow: USSR Academy of Sciences.

Childe, V. G. (1951). *Social evolution*. New York: Henry Schuman.

Chirimuuta, R. C., and Chirimuuta, R. J. (1989). *AIDS, Africa and Racism*, 2d ed. London: Free Association Books.

Chisholm, M. (1982). *Modern world development*. Totowa, N.J.: Barnes and Noble.

Choudhary, A. (1974). *Early medieval village in north-eastern India (A.D. 600–1200)*. Calcutta: Punthi Pustak.

Cipolla, C. (1965). *Guns, sails, and empires: Technological innovation and the early phase of European expansion, 1400–1700*. New York: Pantheon.

Cockburn, A., and Hecht, S. (1989). *The fate of the forest*. London: Verso.

Cohen, I. (1981). Introduction to Weber, M., *General economic history*. New Brunswick, NJ: Transactions.

Cohen, M. (1977). *The food crisis in prehistory*. New Haven: Yale University Press.

Collier, W. F. (1868). *Outlines of general history*. London: T. Nelson and Sons.

Collins, K., and Roberts, D., eds. (1988), *Capacity for work in the tropics*. Cambridge, England: Cambridge University Press.

Collyer, O. (1965). *Birth rates in Latin America*. Berkeley: Institute of International Studies, University of California at Berkeley.

Conklin, H. (1969). Lexicographical treatment of folk taxonomies. In Tyler, ed., *Cognitive anthropology*. New York: Holt, Rinehart, and Winston.

Cooper, J. (1985). In search of agrarian capitalism. In T. Aston, and C. Philpin, eds., *The Brenner debate: Agrarian class structure and economic development in pre-industrial Europe*. Cambridge, England: Cambridge University Press.

Corbridge, S. (1986). *Capitalist world development*. Totowa, NJ: Rowman and Littlefield.

Cordell, D., and Gregory, J. (1987). Introduction to D. Cordell and J. Gregory, eds., *African population and capitalism: Historical perspectives*. Boulder, CO: Westview Press.

Coursey, D. (1967). *Yams*. London: Longmans, Green.

Crone, P. (1989). *Pre-industrial societies*. Oxford: Basil Blackwell.

Croot, P., and Parker, D. (1985). Agrarian class structure and the development of capitalism: France and England compared. In T. Aston, and C. Philpin, eds., *The Brenner debate: Agrarian class structure and economic development in pre-industrial Europe*. Cambridge, England: Cambridge University Press.

Crosby, A. W. (1972). *The Columbian exchange: Biological and cultural consequences of 1492*. Westport, CO: Greenwood Press.

Cross, H. (1983). South American bullion production and export 1550–1750. In J. Richards, ed., *Precious metals and the late medieval and early modern worlds*. Durham, NC: Carolina Academic Press.

Curtin, P. (1969). *The Atlantic slave trade*. Madison: University of Wisconsin Press.

———. (1975). *Economic change in pre-colonial Africa: Senegambia in the era of the slave trade*. Madison: University of Wisconsin Press.

———. (1990). *The rise and fall of the plantation complex*. Cambridge, England: Cambridge University Press.

Dalal, F. (1988). The racism of Jung. *Race and Class* 29(3):1–22.

Darby, H. (1952). *The Domesday geography of eastern England*. Cambridge, England: Cambridge University Press.

Darity, W., Jr. (1990). British industry and the West Indies plantations. *Social Science History* 14(1):117–148.

Das Gupta, A. (1967). *Malabar in Asian trade: 1740–1800*. Cambridge, England: Cambridge University Press.

de Vries, J. (1984). *European urbanization, 1500–1800*. Cambridge, MA: Harvard University Press.

Deerr, Noel. (1949–1950). *The history of sugar*. 2 vols. London: Chapman and Hall.

Denevan, W. (1966), *The aboriginal cultural geography of the llanos de Mojos of Bolivia*. *Iberoamericana*, no. 48. Berkeley and Los Angeles: University of California Press.

———. (1976). Introduction to W. Denevan, ed., *The native population of the Americas in 1492*, 1–12. Madison: University of Wisconsin Press.

———. (1982). Hydraulic agriculture in the American tropics: Forms, measures, and recent research. In K. Flannery, ed., *Maya subsistence: Studies in memory of Dennis E. Puleston*. New York: Academic Press.

———. (1992). The pristine myth: The landscape of the Americas in 1492. *Annals of the Association of American Geographers* 82: 369–385.

———, ed. (1976). *The native population of the Americas in 1492*. Madison: University of Wisconsin Press.

DeVisse J., and Labib, S. (1984). Africa in intercontinental relations. In D. Niane, ed., *UNESCO general history of Africa, Vol. 4*. Paris: UNESCO.

Dew, T. (1853). *A digest of the laws, customs, manners, and institutions of the ancient and modern nations*. New York: Appleton.

Dewey, J. (1916). The logic of judgments of practice. In *Essays in experimental logic*. Chicago: University of Chicago Press.

———. (1929). *The quest for certainty*. New York: G. P. Putnam.

Di Meglio, R. (1970). Arab trade with Indonesia and the Malay Peninsula from the 8th to the 16th century. In D. S. Richards, ed., *Islam and the trade of Asia*. Oxford: Cassirer.

Dirks, N. (1987). *The hollow crown: Ethnohistory of an Indian kingdom*. New York: Cambridge University Press.

Dobb, Maurice. (1947). *Studies in the development of capitalism*. New York: International Publishers.

Doolittle, W. (1990). *Canal irrigation in prehistoric Mexico: The sequence of technological change*. Austin: University of Texas Press.

DuBois, W. E. B. (1965). *The world and Africa: An inquiry into the part which Africa has played in world history*. New York: International Publishers.

Dunn, R. S. (1972). *Sugar and slaves: The rise of the planter class in the English West Indies, 1624–1713*. Chapel Hill: University of North Carolina Press.

Duruy, V., and Grosvenor, E. (1901). *A general history of the world*. New York: Crowell.

Dutt, R. P. (1943). *The problem of India*. New York: International Publishers.

Duyvendak, J. (1933). *Ma Huan re-examined. Verhandlungen der Koninklijke Akademie van Wetenschappen te Amsterdam*, afdeeling letterkunde, nieuwe reeks. Part 33(3):1–74.

Edel, M. (1969). The Brazilian sugar cycle of the 17th century and the rise of West Indian competition. *Caribbean Studies* 9(1):24–45.

Edmonson, M. (1961). Neolithic diffusion rates. *Current Anthropology* 2(2):71–86.

Eliot Smith, G. (1928). *In the beginning: The origin of civilization*. New York: Morrow.

———. (1933). *The diffusion of culture*. London: Watts.

Elvin, M. (1973). *The pattern of the Chinese past*. Stanford, CA: Stanford University Press.

————. (1988). China as a counterfactual. In J. Baechler, J. A. Hall, and M. Mann, eds., *Europe and the rise of capitalism*, 101–112. Oxford: Basil Blackwell.

Engels, F. (1970). *The origin of the family, private property, and the state*. In *Marx and Engels: Selected works*, Vol. 3. Moscow: Progress.

————. (1974). What have the working classes to do with Poland? In D. Fernbach, ed., *Karl Marx: Political writings*, Vol. 3. New York: Vintage.

————. (1975). Letter to K. Marx, June 6, 1853. In *Marx and Engels: Selected correspondence*. Moscow: Progress.

Fei Hsiao-tung. (1953). *China's gentry*. Chicago: University of Chicago Press.

Fiedel, S. (1987). *Prehistory of the Americas*. Cambridge, England: Cambridge University Press.

Filesi, T. (1972). *China and Africa in the Middle Ages*. London: Frank Cass.

Finley, M. (1975). *The use and abuse of history*. New York: Viking Press.

Fisher, D. (1989). The price revolution: A monetary interpretation. *Journal of Economic History* 49:883–902.

Fisher, G. (1885). *Outlines of universal history*. New York: Ivison, Blakeman, Taylor.

————. (1896). *A brief history of the nations and their progress in history*. New York: American Book Co.

Fleck, L. (1979). *Genesis and development of a scientific fact*. Chicago: University of Chicago Press.

Florescano, E. (1984). The formation and economic structure of the hacienda in New Spain. In L. Bethell, ed., *The Cambridge history of Latin America: Colonial Latin America*, Vol. 2. Cambridge, England: Cambridge University Press.

Foster, G. (1962). *Traditional cultures: And the impact of technological change*. New York: Harper and Row.

Frake, C. (1969). The ethnographic study of cognitive systems. In S. Tyler, ed., *Cognitive anthropology*. New York: Holt, Rinehart, and Winston.

Frank, A. G. (1968). *Capitalism and underdevelopment in Latin America*. New York: Monthly Review Press.

————. (1969). Sociology of development and underdevelopment of sociology. In his *Latin America: Underdevelopment or revolution*. New York: Monthly Review Press.

————. (1978). *World accumulation, 1492–1789*. New York: Monthly Review Press.

Frankel, H. (1991). Marxism and physics: A new look. *Science and Society* 55:336–347.

Freedman, M. (1966). *Chinese lineage and society*. London: University of London Press.

Freeman, E. (1874). *General sketch of history*. New York: Holt.

Freire, P. (1972). *Pedagogy of the oppressed*. New York: Herder and Herder.

FRELIMO (Frente de Libertação de Moçambique. (1971). *Historia de Moçambique*. Porto, Portugal: Afrontamento.

Freund, J. (1969). *The sociology of Max Weber*. New York: Vintage.

Fu Chu-fu and Li Ching-neng. (1956). *The sprouts of capitalistic factors within China's feudal society*. Program in East Asian Studies, Western Washington State University, Occasional Paper no. 7.

Furtado, C. (1963). *The economic growth of Brazil*. Berkeley and Los Angeles: University of California Press.

Fyfe, C., and McMaster, D., eds. (1981). *African historical demography*. Edinburgh, Scotland: Centre of African Studies.

Galeano, E. (1972). *The open veins of Latin America*. New York: Monthly Review Press.

Galloway, J. H. (1977). The Mediterranean sugar industry. *Geographical Review* 67:177–192.

———. (1989). *The sugar cane industry: An historical geography from its origins to 1914.* Cambridge, England: Cambridge University Press.

Garraty, J., and Gay, P., eds. (1981). *The Columbia history of the world.* New York: Dorset Press.

Gernet, J. (1962). *Daily life in China on the eve of the Mongol invasion, 1250–1276.* Stanford, CA: Stanford University Press.

Giblin, J. (1990). Trypanosomiasis control in African history: An evaded issue? *Journal of African History* 31:59–80.

Gilfillan, S. (1920). The coldward course of progress. *Political Science Quarterly* 35:393–410.

Gill, D. and Levidow, L., eds. (1987). *Anti-racist science teaching.* London: Free Association Books.

Gilman, A. (1874). *First steps in general history.* Cambridge, MA: Riverside Press.

Godelier, M. (1969). *Sobre el modo de producción asiático.* Barcelona, Spain: Ediciones Martínez Roca.

Goitein, S. (1967). *A Mediterranean society.* Berkeley and Los Angeles: University of California Press.

Golson, J. (1977). No room at the top: Agricultural intensification in the New Guinea Highlands. In J. Allen et al., eds., *Sunda and Sahul: Prehistoric studies in Southeast Asia, Melanesia and Australia.* London: Academic Press.

Goodrich, S. G. (1843). *Peter Parley's common school history.* Philadelphia: E. H. Butler.

Gopal, L. (1963). Quasi-manorial rights in ancient India. *Journal of the Economic and Social History of the Orient* 6:296–308.

Gopal, S. (1972). Nobility and the mercantile community in India. *Journal of Indian History* 50:793–802;

Gorman, C. (1977). A *priori* models and Thai prehistory: A reconsideration of the beginnings of agriculture in Southeast Asia. In C. A. Reed, ed., *Origins of agriculture.* The Hague: Mouton.

Gossett, T. (1963). *Race: The history of an idea in America.* Dallas: Southern Methodist University Press.

Gould, P. (1969). *Spatial diffusion.* Commission on College Geography, Association of American Geographers, Resource Paper no. 4, Washington, DC: Association of American Geographers.

Graebner, F. (1911). *Methode der ethnologie.* Heidelberg, Germany.

Gupta, N. S. (1970). *Industrial structure of India during the medieval period.* Delhi: Chand.

Habib, I. (1963). *The agrarian system of Mughal India.* London: Asia Publishing House.

———. (1964). Usury in medieval India. *Comparative Studies in Society and History* 6:392–419.

———. (1965). The social distribution of landed property in pre-British India. *Enquiry* 2(2,n.s.):21–75.

———. (1969). Problems of Marxist historical analysis. *Enquiry* 3(2,n.s.):52–67.

Hagen, E. (1962). *The theory of social change: How economic growth begins.* Homewood, IL: Dorsey Press.

———. (1962). A framework for analyzing economic and political change. In Brookings Institution, ed., *Development of the emerging countries: An agenda for research,* 1–39. Washington, DC: Brookings Institution.

Hägerstrand, T. (1967). *Innovation diffusion as a spatial process*. Translated by A. Pred. Chicago: University of Chicago Press.

Hajnal, J. (1965). European marriage patterns in perspective. In D. Glass and D. Eversley, eds., *Population in history: Essays in historical demography*. Chicago: Aldine.

Hall, J. (1985). *Powers and liberties: The causes and consequences of the rise of the West*. Oxford: Basil Blackwell.

———. (1988). States and societies: The miracle in historical perspective. In J. Baechler, J. A. Hall, and M. Mann, eds., *Europe and the rise of capitalism*. Oxford: Basil Blackwell.

Hallam, H. (1975). The medieval social picture. In E. Kamenka and R. Neale, eds., *Feudalism, capitalism and beyond*. London: Edward Arnold.

Haller, J. (1971). *Outcasts from evolution: Scientific attitudes of racial inferiority, 1859–1900)*. Urbana: University of Illinois Press.

Hamilton, E. J. (1929). American treasure and the rise of capitalism. *Economica* 9:338–357.

———. (1934). *American treasure and the price revolution in Spain, 1501–1650*. Cambridge, MA: Harvard University Press.

Hamilton, G. S. (1985). Why no capitalism in China? Negative questions in historical, comparative research. *Journal of Developing Societies* 1:187–211.

Handler, R. (1989). Review of Macfarlane, A., *Marriage and love in England*. *American Anthropologist* 91:1078–1079.

Hannemann, M. (1975). *The diffusion of the Reformation in Southwestern Germany, 1518–1534*. Department of Geography, University of Chicago, Research Paper no. 167.

Hansis, R. (1976). *Ethnogeography and science: Viticulture in Argentina*. Ph.D. diss., Pennsylvania State University.

Hardy, F. (1936). Some aspects of tropical soils. *Transactions of the Third International Congress of Soil Science* 2:150–163.

Harewood, J. (1966). Population growth in Grenada. *Social and Economic Studies* 15:61–84.

Harris, M. (1968). *The rise of anthropological theory*. New York: Crowell.

Harrison, P. (1969). *The communists and Chinese peasant rebellions*. New York: Atheneum.

———, and Turner, B. L., eds. (1978). *Pre-Hispanic Maya agriculture*. Albuquerque: University of New Mexico Press.

Hasan, A. (1969). The silver currency output of the Mughal Empire and prices in India during the 16th and 17th centuries. *Indian Economic and Social History Review* 6:85–116.

Haskel, D. (1848). *Chronology and universal history*. New York: J. H. Colton.

Hassan, F. (1978). Demographic archeology. In M. B. Schaffer, ed., *Advances in archeological method and theory*, Vol. 1. New York: Columbia University Press.

Hegel, G. (1956). *The philosophy of history*. New York: Dover Publications.

Hilton, R. (1980). Individualism and the English peasantry. *New Left Review*, no. 120, pp. 109–111.

———. (1985). Introduction. In T. Aston, and C. Philpin, eds., *The Brenner debate: Agrarian class structure and economic development in pre-industrial Europe*. Cambridge, England: Cambridge University Press.

———, ed. (1976). *The transition from feudalism to capitalism*. London: New Left Books.

Ho Ping-ti. (1977). The indigenous origins of Chinese agriculture. In C. A. Reed, ed., *Origins of agriculture*. The Hague: Mouton.

Hodgett, A. J. (1972). *A social and economic history of medieval Europe*. London: Methuen.

Hopkins, A. (1973). *An economic history of West Africa*. New York: Columbia University Press.

Howard, A. (1975). Pre-colonial centers and regional systems in Africa. *Pan-African Journal* 8(3):247–270.

Huddleston, L. (1967). *Origins of the American Indians: European concepts, 1492–1729*. Austin: University of Texas Press.

Hoyle, R. (1990). Tenure and the land market in early modern England: Or a late contribution to the Brenner debate. *Economic History Review* 43:1–20.

Hudson, B. (1977). The new geography and the new imperialism: 1870–1918. *Antipode* 9(1):12–19.

Hulme, P. (1992). *Colonial encounters: Europe and the Native Caribbean 1492–1797*. London: Routledge.

Huntington, E. (1924). *Principles of human geography*. 3d ed. New York: John Wiley.

Inikori, J. (1987). Slavery and the development of industrial capitalism. In B. Solow and S. Engerman, eds. (1987). *British capitalism and Caribbean slavery: The legacy of Eric Williams*. Cambridge, England: Cambridge University Press.

———. (1989). Slavery and the revolution in cotton textile production in England. *Social Science History* 13(4):343–379.

———, ed. (1979). *The African slave trade from the fifteenth to the nineteenth century*. Paris: UNESCO.

Irvine, F. (1934). *A textbook of West African agriculture: Soils and crops*. London: Oxford University Press.

Irwin, G. (1982). Sub-Saharan Africa. In J. Garraty, and P. Gay, eds., *Columbia history of the world*. New York: Columbia University Press.

Isichei, E. (1982). *A history of Nigeria*. London: Longmans.

Ivanov, V. (1985). Round-table: State and law in the ancient Orient. (Discussion.) *Social Sciences* (USSR Academy of Sciences) 16(3):177–201.

Jackson, P., ed. (1987). *Race and racism: Essays in social geography*. London: Allen and Unwin.

James, C. L. R. (1938a). *The Black Jacobins: Toussaint L'Ouverture and the San Domingo revolution*. London: Secker and Warburg.

———. (1938b). *A history of Negro revolt*. London: Fact. Rev. ed. (1969), *A history of Pan-African revolt*. Washington, DC: Drum and Spear Press.

———. (1970). The Atlantic slave trade and slavery: Some interpretations of their significance in the development of the United States and the Western World. In J. A. Williams and C. F. Harris, eds, *Amistad I*, 119–164. New York: Vintage.

Jett, S. C. (1991). Further information on the geography of the blowgun and its implications for transoceanic contacts. *Annals of the Association of American Geographers* 81:89–102.

Jha, S. C. (1963). *Studies in the development of capitalism in India*. Calcutta: Mukhopadhyay.

Johnson, K. (1977). *Do as the land bids: A study of Otomí resource use on the eve of irrigation*. Ph.D. diss., Clark University.

Joly, N. (1897). *Man before metals*. New York: Appleton.

Jones, E. L. (1981). *The European miracle*. Cambridge, England: Cambridge University Press.

——. (1988). *Growth recurring: Economic change in world history*. Oxford: Clarendon Press.

Jung, C. (1963). *Memories, dreams, reflections*. New York: Random House.

——. (1964). The dreamlike world of India. In his *Civilization in transition* (*Collected works*, vol. 10). Princeton, NJ: Princeton University Press.

——. (1971). *Psychological types* (*Collected works*, vol. 6). Princeton, NJ: Princeton University Press.

Kabaker, A. (1977). A radiocarbon chronology relevant to the origins of agriculture. In C. A. Reed, ed., *Origins of agriculture*. The Hague: Mouton.

Kagan, D., Ozment, S., and Turner, F. (1987). *The Western heritage*, 2 vols., 3rd ed. New York: Macmillan.

Kea, R. (1982). *Settlements, trade, and polities in the seventeenth-century Gold Coast*. Baltimore: Johns Hopkins University Press.

Keightley, T. (1849). *Outlines of history*. Rev. ed. London: Longmans.

Kerridge, E. (1968). *The agricultural revolution*. New York: A. M. Kelley.

Kertzer, D. (1989). The joint family household revisited: Demographic constraints and household complexity in the European past. *Journal of Family History* 14:1–16.

Kinder, H., and Hilgemann, W. (1974). *The Anchor atlas of world history*, Vol. 1. New York: Anchor Books.

Kitching, G. (1983). Proto-industrialization and demographic change. *Journal of African History* 24:221–240.

Kniffen, F. (1965). Folk housing: Key to diffusion. *Annals of the Association of American Geographers* 55:549–577.

Koepping, K.-P. (1983). *Adolf Bastian and the psychic unity of mankind*. St. Lucia, Australia: University of Queensland Press.

Kosambi, D. D. (1969). *Ancient India*. New York: Meridian Books.

Kroeber, A. (1937). Diffusionism. In *Encyclopedia of the social sciences*, 5:139–142. New York: Macmillan.

Kuhn, T. (1970). *The structure of scientific revolutions*. 2nd ed. Chicago: University of Chicago Press.

Laclau, E. (1977). *Politics and ideology in Marxist theory*. London: New Left Books.

Laibman, D. (1984). Modes of production and theories of transition. *Science and Society* 48(3):257–294.

Lambert, D. (1971). The role of climate in the economic development of the tropics. *Land Economics* 47:339–344.

Lane, F. (1973). *Venice: A maritime republic*. Baltimore: Johns Hopkins University Press.

Laslett, P. (1988). The European family and early industrialization. In J. Baechler, J. A. Hall, and M. Mann, eds., *Europe and the rise of capitalism*. Oxford: Basil Blackwell.

Lattimore, O. (1980). The periphery as locus of innovation. In J. Gottmann, ed., *Centre and periphery: Spatial variation and politics*. Beverly Hills. CA: Sage Publications.

Lee, G. (1987). Comparative perspectives. In M. B. Sussman and S. K. Steinmetz, eds., *Handbook of marriage and the family*. New York: Plenum.

Lee, R. (1992). Art, science, or politics? The crisis in hunter-gatherer studies. *American Anthropologist* 94:31–54.

Lelekov, L. (1985). Round-table: State and law in the ancient Orient. (Discussion.) *Social Sciences* (USSR Academy of Sciences) 16(3):177–201.

Lentnek, B. (1969). Economic transition from traditional to commercial agriculture. *Annals of the Association of American Geographers* 59:65–84.

Lerner, R., Meacham, S., and Burns, E. (1988). *Western civilizations: Their history and their culture.* 11th ed. New York: Norton.

Le Roy Ladurie, E. (1985). A reply to Robert Brenner. In T. Aston and C. Philpin, eds., *The Brenner debate: Agrarian class structure and economic development in pre-industrial Europe.* Cambridge, England: Cambridge University Press.

Levitzion, N. (1972). The early states of the Western Sudan to 1500. In J. Ajayi and M. Crowder, eds., *History of West Africa*, Vol. 1. New York: Columbia University Press.

Lévy-Bruhl, L. (1966). *How natives think.* New York: Washington Square Press.

Lewis, A. (1973). Maritime skills in the Indian Ocean, 1368–1500. *Journal of the Economic and Social History of the Orient* 16:238–264.

Lewis, W. A., ed. (1970). *Tropical development, 1880–1913.* Evanston, IL: Northwestern University Press.

Liceria, M. A. C. (1974). Emergence of Brahmanas as landed intermediaries in Karnataka, c. A.D. 1000–1300. *Indian Historical Review* 1(1):28–35.

Lloyd, G. (1990). *Demystifying mentalities.* Cambridge, England: Cambridge University Press.

Lo Jung-pang. (1955). China as a sea power. *Far Eastern Quarterly* 14:489–503.

Lopez, R. (1953). Hard times and investment in culture. In R. Lopez, ed., *The Renaissance: A symposium.* New York: Metropolitan Museum of Art.

———, and Miskimin, H. (1961–1962). The economic depression of the Renaissance. *Economic History Review* 14:408–426.

Lord, J. (1869). *Ancient states and empires: For colleges and schools.* New York: Charles Scribner.

Lovell, W. G. (1992). "Heavy shadows and black night": Disease and depopulation in Colonial Spanish America. *Annals of the Association of American Geographers* 82: 426–443.2

Lowie, R. (1937). *The history of ethnological theory.* New York: Rinehart.

Löwith, K. (1982). *Max Weber and Karl Marx.* London: Allen and Unwin.

Ma Huan. (1970). *The overall survey of the ocean's shores.* Cambridge, England: Cambridge University Press.

Mabogunji, A. (1972). The land and peoples of West Africa. In J. Ajayi and M. Crowder, eds., *History of West Africa*, Vol. 1. New York: Columbia University Press.

Macauley, T. (1850–1861). *The history of England from the accession of James the Second.* 5 vols. London: Longman.

Macfarlane, A. (1978). *The origins of English individualism.* Oxford: Basil Blackwell.

———. (1986). *Marriage and love in England: Modes of reproduction 1300–1840.* Oxford: Basil Blackwell.

———. (1988) The cradle of capitalism. In J. Baechler, J. A. Hall, and M. Mann, eds., *Europe and the rise of capitalism.* Oxford: Basil Blackwell.

Magalhães-Godinho, V. (1969). *L'economie de l'empire portugais aux XV et XVI siècles.* Paris: SEVPEN.

Magubane, B. (1987). *The ties that bind: African-American consciousness and Africa.* Trenton, NJ: Africa World Press.

Mahalingam, T. (1951). *Economic life in the Vijayanagar Empire.* Madras: University of Madras Press.

Malinowski, B. (1927). The life of culture. In G. Eliot Smith, ed., *Culture: The diffusion controversy.* New York: Norton.

Mamdani, M. (1972). *The myth of population control.* New York: Monthly Review Press.

Mandelbaum, M. (1971). *History, man and reason: A study in nineteenth-century thought.* Baltimore: Johns Hopkins University Press.

Mann, M. (1986). *The sources of social power: Vol. 1. A history of power from the beginning to A.D. 1760.* Cambridge, England: Cambridge University Press.

———. (1988). European development: Approaching a historical explanation. In J. Baechler, J. A. Hall, and M. Mann, eds., *Europe and the rise of capitalism.* Oxford: Basil Blackwell.

Mannheim, K. 1936. *Ideology and utopia.* New York: Harcourt, Brace.

Marglin, S. (1990). Losing touch: The cultural conditions of worker accommodation and resistance. In F. Marglin and S. Marglin, eds., *Dominating knowledge: Development, culture, and resistance.* Oxford: Clarendon Press.

Markham, S. (1944), *Climate and the energy of nations.* London: Oxford University Press.

Marx, K. (1976). *Capital.* New York: Vintage.

———. (1979). The British rule in India. In *Marx and Engels: Collected Works,* 12:125–33. Moscow: Progress.

———, and Engels, F. (1975). *Selected correspondence.* 3d ed. Moscow: Progress.

———, and Engels, F. (1976). *The German ideology.* Moscow: Progress.

McAlister, L. (1984). *Spain and Portugal in the New World, 1492–1700.* Minneapolis: University of Minnesota Press

McClelland, D. (1961). *The achieving society.* Princeton, NJ: Van Nostrand.

McKay, D. (1943). Colonialism in the French geographical movement. *Geographical Review* 33:214–232.

McNeill, W. (1967). *A world history.* New York: Oxford University Press.

———. (1976). *Plagues and peoples.* Garden City, NY: Anchor Books.

Mead, G. H. (1936). *Movements of thought in the nineteenth century.* Chicago: University of Chicago Press.

———. (1938). *The philosophy of the act.* Chicago: University of Chicago Press.

Mead, M. (1930). *Growing up in New Guinea.* New York: Blue Ribbon Books.

Megaw, J., ed. (1977). *Hunters, gatherers and first farmers beyond Europe: An archeological survey.* Leicester, England: Leicester University Press.

Merriman, T. (1875). *The trail of history.* 3d ed. Boston: Merriman and Stewart.

Miller, J. (1982). The significance of drought, disease, and famine in the agriculturally marginal zones of West-Central Africa. *Journal of African History* 23:17–61.

———. (1988). *Way of death.* Madison: University of Wisconsin Press.

Milne, G. (1947). A soil reconnaissance journey through parts of Tanganyika Territory, December 1935 to February 1936. *Journal of Ecology* 35:192–265.

Minchinton, W. (1969). *The growth of English overseas trade.* London: Methuen.

Mintz, S. W. (1985). *Sweetness and power: The place of sugar in modern history.* New York: Penguin Books.

Miskimin, H. (1969). *The economy of early Renaissance Europe, 1300–1460.* Englewood Cliffs, NJ: Prentice-Hall.

Mohr, E., and van Baren, F. (1954). *Tropical soils*. Amsterdam: Royal Tropical Institute.

Montesquieu. (1949). *The spirit of the laws*. New York: Hafner Press. (Originally published in 1748.)

Mukherjee, Radhakamal. (1967). *The economic history of India, 1600–1800*. Allahabad, India: Kitab Mahal.

Mukherjee, Ramkrishna. (1958). *The rise and fall of the East India Company*. 2d ed. Berlin: VEB Deutscher Verlag der Wissenschaften.

Müller J. (1842). *The history of the world to 1783*. 4 vols. Boston: Webb.

Nag, M. (1980). How modernization can also increase fertility. *Current Anthropology* 21:571–588.

Naqvi, H. K. (1968). *Urban centres in Upper India, 1556–1803*. Bombay: Asia Publishing House.

Needham, J. (1985). *Gunpowder as the fourth estate east and west*. Hong Kong: Hong Kong University Press.

———, and collaborators. (1954–1984). *Science and civilization in China*. 6 vols. Cambridge, England: Cambridge University Press.

Neumann, E. (1954). *The origins and history of consciousness*. New York: Scribner's.

Newsom, L. (1985). Indian population patterns in colonial Spanish America. *Latin American Research Review* 20(3): 41–74.

Niane, D. (1984). Mali and the second Mandingo expansion. In D. Niane, ed., *UNESCO general history of Africa*, Vol. 4. Paris: UNESCO.

———, ed. (1984). *UNESCO general history of Africa*, Vol. 4. Paris: UNESCO.

Nicholas, D. M. (1967–68). Town and countryside: Social and economic tensions in 14th century Flanders. *Comparative Studies in Society and History* 10:458–485.

Nisbet, R. (1980). *History of the idea of progress*. New York: Basic Books.

Nurul Hasan, S. (1969). The position of the Zamindars in the Mughal Empire. In R. Frykenberg, ed., *Land control and social structure in Indian history*. Madison: University of Wisconsin Press.

Nye, P., and Greenland, D. (1960). *The soil under shifting cultivation*. Farnham Royal, England: Commonwealth Agricultural Bureaux.

O'Keefe, P., and Wisner, B. (1975). African drought: The state of the game. In P. Richards, ed., *African environment: Problems and perspectives*, London: International African Institute.

Onimode, B. (1982). *Imperialism and underdevelopment in Nigeria*. London: Zed Books.

Orwin, C., and Orwin, C. (1967). *The open fields*. Oxford: Clarendon Press.

Osae, T., Nwabara, S., and Odunsi, A. (1973). *A short history of West Africa*. New York: Hill and Wang.

Pacey, A. (1990). *Technology in world history: A thousand-year history*. Cambridge, MA: MIT Press.

Padgug, R. (1976). Problems in the theory of slavery and slave society. *Science and Society* 40:3–28.

Palmer, R., ed. (1957). *Atlas of world history*. Chicago: Rand McNally.

Panikkar, K. M. (1959). *Asia and Western influence*. London: Allen and Unwin.

Parsons, J. B. (1970). *Peasant rebellions of the late Ming Dynasty*. Tucson: University of Arizona Press.

Pearse, A. (1980). *Seeds of plenty, seeds of want: Social and economic implications of the green revolution*. Oxford: Clarendon Press.

Pendleton, R. (1943). Land use in north-eastern Thailand. *Geographical Review* 33:14–41.

Perry, W. (1935). *The primordial ocean.* London: Methuen.

Piaget, J. (1971). *Psychology and epistemology.* New York: Grossman Publishers.

Pinar, W. (1974). *Heightened consciousness, cultural revolution, and curriculum theory.* Berkeley, CA: McCutchan.

Pires, T. (1944). *The Suma Oriental.* London: Hakluyt Society.

Ploetz, C., and Tillinghast, W. (1883). *Epitome of ancient, medieval, and modern history.* Boston: Houghton-Mifflin.

Prakash, I. (1964). Organization of industrial production in urban centres in India during the 17th century with special reference to textiles. In B. Ganguli, ed., *Readings in Indian economic history.* Bombay, India.

Prescott, J., and Pendleton, R. (1952). *Laterites and lateritic soils.* Slough, England: Commonwealth Agricultural Bureaux.

Purcell, V. (1965). *The Chinese in Southeast Asia.* 2d ed. London: Oxford University Press.

Qaisar, A. J. (1974). The role of brokers in medieval India. *Indian Historical Review* 1:220–246.

Quackenbos, J. (1889). *Illustrated school history of the world.* New York: American Book Company.

Radell, D. R. (1976). The Indian slave trade and population of Nicaragua during the sixteenth century. In W. Denevan, ed., *The Native Population of the Americas in 1492.* Madison: University of Wisconsin Press.

Radin, P. (1927). *Primitive man as philosopher.* New York: Appleton.

———. (1965). *The method and theory of ethnology.* New York: Basic Books.

Ratzel, F. (1896). *The history of mankind (Völkerkunde).* 2 vols. London: Macmillan.

Rawski, E. (1972). *Agricultural change and the peasant economy of South China.* Cambridge, MA: MIT Press.

Raychaudhuri, T. (1962). *Jan Company in Coromandel.* The Hague: Nijhoff.

———. (1965). The agrarian system of Mughal India. *Enquiry* 2(1,n.s.):92–121.

Reclus, E. (1876–1894) *Nouvelle géographie universelle.* Paris: Hachette.

Reed, C. A., ed. (1977). *Origins of agriculture.* The Hague: Mouton.

Richards, P., ed. (1975). *African environment: Problems and perspectives.* London: International African Institute.

Robbins, R. (1832). *The world displayed in its history and geography: Embracing a history of the world from the Creation to the present day.* New York: W. W. Reed.

Roberts, J. (1987). *The Hutchinson history of the world.* 2d ed. London: Hutchinson.

Robinson, C. (1987). Capitalism, slavery and bourgeois historiography. *History Workshop* 23:122–141.

Rodinson, M. (1973). *Islam and capitalism.* New York: Pantheon.

———. (1974). Le marchand musulman. In D. S. Richards, ed., *Islam and the trade of Asia.* Cambridge, England: Cassirer.

Rodney, W. (1970). *A history of the Upper Guinea Coast, 1545–1800.* New York: Monthly Review Press.

———. (1972). *How Europe underdeveloped Africa.* London: Bogle-L'Ouverture Publications, and Dar es Salaaam: Tanzania Publishing House.

Rogers, E. (1962). *The diffusion of innovations.* New York: Free Press.

———, and Shoemaker, F. (1971). *Communication of innovations.* New York: Free Press.

Rorty, R. (1980). *Philosophy and the mirror of nature.* Oxford: Basil Blackwell.

Rouse, I. (1961). Comments on Edmonson's Neolithic diffusion rates. *Current Anthropology* 2:96.

————. (1986). *Migrations in prehistory.* New Haven: Yale University Press.

Roxborough, I. (1979). *Theories of underdevelopment.* London: Macmillan.

Sack, R. (1980). *Conceptions of space in social thought.* London: Macmillan.

Said, E. (1979). *Orientalism.* New York: Random House.

Sanderson, E. (1898). *History of the world from the earliest time to the year 1898.* New York: Appleton.

Sastri, K. N. (1966). *A history of South India from prehistoric times to the fall of Vijayanagar.* 3d ed. Madras: Oxford University Press.

Schmidt, W. (1939). *The culture historical method of ethnology.* New York: Fortuny.

Schweder, R. (1990). Cultural psychology: What is it? In J. Stigler, R. Schweder, and G. Herdt, eds., *Cultural psychology: Essays on comparative human development.* Chicago: University of Chicago Press.

Seccombe, W. (1983). Marxism and demography. *New Left Review* no. 137, pp. 22–47.

Semo, E. (1982). *Historia del capitalismo en México: Los orígenes, 1521–1763.* Mexico, D.F.: Ediciones Era.

Shannon, G., and Pyle, G. (1989). The origin and diffusion of AIDS. *Annals of the Association of American Geographers* 79:1–24.

Sharma, R. S. (1965). *Indian feudalism, c.300–1200.* Calcutta: University of Calcutta Press.

————. (1966). *Light on early Indian society and economy* Bombay: Manaktalas.

Shepherd, W. (1911). *Historical atlas.* New York: Henry Holt.

Sheridan, R. (1974). *Sugar and slavery.* Baltimore, MD: Johns Hopkins University Press.

————. (1987). Eric Williams and *Capitalism and slavery:* A biographical and historiographical essay. In B. Solow and S. Engerman, eds. *British capitalism and Caribbean slavery: The legacy of Eric Williams.* Cambridge, England: Cambridge University Press.

Sherif, A. (1987). *Slaves, spices and ivory in Zanzibar.* London: James Currey.

Simkin C., (1968). *The traditional trade of Asia.* London: Oxford University Press.

Simonsen, R. (1944). *História econômica do Brasil, 1500–1820,* 2d ed. São Paulo, Brazil: Companhia Editora Nacional.

Sinclair, P. (1991). Archeology in eastern Africa: An overview of current chronological issues. *Journal of African History* 32:179–219.

Smith, A. (1972). The early states of the Central Sudan. In J. Ajayi and M. Crowder, eds., *History of West Africa,* Vol. 1. New York: Columbia University Press.

Smith, C. T. (1969). *An historical geography of Western Europe before 1800.* London: Longmans.

Smith, P. (1868). *An ancient history: From the earliest records to the fall of the Western Empire.* 3 vols. London: Walton.

So Kwan-wai. (1975). *Japanese piracy in Ming China during the 16th century.* East Lansing: Michigan State University Press.

Solow, B. (1987). Capitalism and slavery in the exceedingly long run. In B. Solow and S. Engerman, eds. *British capitalism and Caribbean slavery: The legacy of Eric Williams.* Cambridge, England: Cambridge University Press.

————, and Engerman, S., eds. (1987). *British capitalism and Caribbean slavery: The legacy of Eric Williams.* Cambridge, England: Cambridge University Press.

Spencer, H. (1969). *The man versus the state.* Baltimore: Penguin Books.

Spradley, J., and McCurdy, D. (1975). *Anthropology: A cultural perspective.* New York: Wiley.

Steele, J., and Steele, E. (1883). *A brief history of ancient, medieval, and modern peoples.* New York: American Book Co.

Steward, J. (1955). *Theory of culture change: The methodology of multilinear evolution.* Urbana: University of Illinois Press.

Stocking, G. (1968). *Race, culture, and evolution.* New York: Free Press.

———. (1987). *Victorian anthropology.* New York: Free Press.

———, ed. (1984). *Functionalism historicized: Essays on British social anthropology.* Madison: University of Wisconsin Press.

Stone, L. (1977). *The family, sex and marriage in England 1500–1800.* New York: Harper and Row.

Sweezy, P. (1976). A critique. In R. Hilton, ed., *The transition from feudalism to capitalism.* London: New Left Books.

Swindell K. (1981). Domestic production, labor mobility, and population change in West Africa, 1900–1980. In C. Fyfe and D. McMaster, eds., *African historical demography.* Edinburgh, Scotland: Centre of African Studies.

Swinton, W. (1874). *Outlines of the world's history.* New York: Ivison, Blakeman, Taylor.

Taeuber, I. (1970). The families of Chinese farmers. In M. Freedman, ed., *Family and kinship in Chinese society.* Stanford, CA: Stanford University Press.

Talkington, L. (1986). But the editor looks at the universe from a different frame of reference. *Science and Nature*, nos. 7–8, pp. 94–100.

Tarde, G. (1903). *The laws of imitation.* New York: Henry Holt.

Tawney, R. H. (1952). *Religion and the rise of capitalism.* New York: Harcourt Brace.

Taylor, G. (1945). *Environment and nation.* Chicago: University of Chicago Press.

Temu, A., and Swai, B. (1981). *Historians and Africanist history: A critique.* London: Zed Books.

Thalheimer, M. (1883). *Outline of general history for the use of schools.* New York: American Book Co.

Thapar, R. (1978). *Ancient Indian social history: Some interpretations.* New Delhi, India: Orient Longman.

———. (1982). Ideology and the interpretation of early Indian history. *Review* 5:389–412.

Thorndyke, L. (1943). Renaissance or prenaissance? *Journal of the History of Ideas* 4:65–74.

Thrupp, S. (1948). *The merchant class of medieval London (1300–1500).* Ann Arbor: University of Michigan Press.

———. (1972). Medieval industry 1000–1500. In C. Cipolla, ed., *The Fontana economic history of Europe: The Middle Ages.* London: Collins.

Tiedemann, C., and Van Doren, C. (1964). *The diffusion of hybrid seed corn in Iowa: A spatial diffusion study.* Institute for Community Development and Services, Michigan State University, Bulletin B-44.

Titow, J. Z. (1969). *English rural society, 1200–1350.* London: Allen and Unwin.

Tolman, E. (1951). A Psychological model. In T. Parsons and E. Shils, eds., *Towards a general theory of action.* Cambridge, MA: Harvard University Press.

Torras, J. (1980). Class struggle in Catalonia. *Review* 4:253–265.

Toulmin, S., and Goodfield, J. (1965). *The discovery of time.* New York: Harper and Row.

Toyoda, T. (1969). *History of pre-Meiji commerce in Japan.* Tokyo: Japan Cultural Society.

Trigger, B. (1989). *A history of archeological thought*. Cambridge, England: Cambridge University Press.

Tung, C. (1979). *Outline history of China*. Hong Kong: Joint Publishing Co.

Turner, B. (1978). *Marx and the end of Orientalism*. London: Allen and Unwin.

Turshen, M. (1987). Population growth and the deterioration of health: Mainland Tanzania, 1920–1960. In D. Cordell and J. Gregory, eds., *African population and capitalism: Historical perspectives*. Boulder, CO: Westview Press.

Tyler, S., ed. (1969). *Cognitive anthropology*. New York: Holt, Rinehart, and Winston.

Tytler, A. (1844). *Universal history, from the Creation to the beginning of the eighteenth century*. Rev. ed. Boston: Massey.

Udovitch, A. (1974). Commercial techniques in early medieval Islamic trade. In D. S. Richards, ed., *Islam and the trade of Asia*. Cambridge, England: Cassirer.

Usman, Y. B. (1981). *The transformation of Katsina (1400–1883)*. Zaria, Nigeria: Ahmadu Bello University Press.

Van Leur, J. C. (1955). *Indonesian trade and society*. The Hague: W. van Hoeve.

Van Sertima, I. (1976). *They came before Columbus*. New York: Random House.

Van Young, E. (1983). Mexican rural history since Chevalier: The historiography of the colonial hacienda. *Latin American Research Review* 28(3):5–61.

Vavilov, N. (1951). *The origin, variation, immunity, and breeding of cultivated plants*. New York: Ronald Press.

Venturi, F. (1963). The history of the concept of "oriental despotism" in Europe. *The Journal of the History of Ideas* 24:133–143.

Vilar, P. (1976). *A history of gold and money, 1450–1920*. London: New Left Books.

Vishnu-Mittre. (1978). Origin and history of agriculture in the Indian subcontinent. *Journal of Human Evolution* 7:31–36.

Vives, J. Vicens. (1969). *An economic history of Spain*. Princeton: Princeton University Press.

Wachtel, N. (1984). The Indian and Spanish conquest. In Bethell, L., ed., *The Cambridge history of Latin America: Colonial Latin America*, Vol. 1. Cambridge, England: Cambridge University Press.

Wai Andah, B. (1981). West Africa before the seventh century. In G. Mokhtar, ed., *UNESCO general history of Africa: Vol. 2. Ancient civilizations of Africa*. Paris: UNESCO.

Wallerstein, I. (1974–1988). *The modern world system*. 3 vols. New York: Academic Press.

Warren, B. (1980). *Imperialism: Pioneer of capitalism*. London: New Left Books.

Watson, A. (1983). *Agricultural innovation in the early Islamic world: The diffusion of crops and farming techniques, 700–1100*. Cambridge, England: Cambridge University Press.

Watts, S. J., Okello, R., and Watts, S. (1990). Medical geography and AIDS. *Annals of the Association of American Geographers* 80:301–304.

Webb, W. P. (1951). *The great frontier*. Austin: University of Texas Press.

Weber, G. (1853). *Outlines of universal history*. Rev. ed. Edited by F. Brown. Boston: Brewer and Tileston.

Weber, M. (1951). *The religion of China*. New York: Free Press.

———. (1958). *The Protestant ethic and the spirit of capitalism*. New York: Scribner's.

———. (1967). *The religion of India*. New York: Free Press.

———. (1968). *Economy and society*. 2 vols. New York: Bedminster Press.

———. (1976). *The agrarian sociology of ancient civilizations*. London: New Left Books.

———. (1981). *General economic history*. New Brunswick, NJ: Transaction Books.

Wendorf, F., and Schild, R. (1980). *Prehistory of the Eastern Sahara*. New York: Academic Press.

Werner, K. F. (1988). Political and social structures of the West, 300–1300. In J. Baechler, J. A. Hall, and M. Mann, eds., *Europe and the rise of capitalism*. Oxford: Basil Blackwell.

Werner, H., and Kaplan, B. (1964). *Symbol formation*. New York: John Wiley.

Wheatley, P. (1961). *The Golden Khersonese: Studies in the historical geography of the Malay Peninsula before A.D. 1500*. Kuala Lumpur: University of Malaya Press.

———. (1971). *The pivot of the four quarters: A preliminary inquiry into the origins and character of the ancient Chinese city*. Chicago: Aldine.

Whelpley, S. (1844). *A compound of history, from the earliest times*. 12th ed. New York: Collins Brothers.

White, G., ed. (1974). *Natural hazards: Local, national, global*. New York: Oxford University Press.

White, Lynn, Jr. (1962). *Medieval technology and social change*. London: Oxford University Press.

———. (1968). *Machina ex deo: Essays in the dynamism of Western culture*. Cambridge, MA: MIT Press.

———. (1968) The historical roots of our ecological crisis. In his *Machina ex deo: Essays in the dynamism of Western culture*, 75–94. Cambridge, MA: MIT Press.

Whitehead, A. N. (1929). *Process and reality*. New York: Humanities Press.

———. (1938). *Modes of thought*. New York: Macmillan.

———. (1948). *Science and philosophy*. New York: Philosophical Library.

Whitman, J. (1984). Philology to anthropology in mid-nineteenth-century Germany. In G. Stocking, ed., *Functionalism historicized: Essays on British social anthropology*. Madison: University of Wisconsin Press.

Whitmore, T. (1991). A simulation of sixteenth-century population collapse in the Basin of Mexico. *Annals of the Association of American Geographers* 81:464–487.

Wiethoff, B. (1963). *Die Chinesische Seeverbotspolitik und der Private Überseehandel von 1368 bis 1567*. Hamburg, Germany: Gesellschaft für Natur- und Völkerkunde Ostasiens.

Wilken, G. (1987). *Good farmers: Traditional resource management in Mexico and Central America*. Berkeley and Los Angeles: University of California Press.

Willard, E. (1845). *Universal history in perspective*. 2d ed. New York: Barnes.

Williams, E. (1944). *Capitalism and slavery*. Chapel Hill: University of North Carolina Press.

———. (1966). *British historians and the West Indies*. London: Andre Deutsch.

Williams, R. (1990). *The American Indian in Western legal thought: The discourses of conquest*. New York: Oxford University Press.

Wisner, B. (1989). *Power and need in Africa*. Trenton, NJ: Africa World Press.

———, and Mbithi, P. (1974). Drought in Eastern Kenya. In G. White, ed., *Natural hazards: Local, national, global*. New York: Oxford University Press.

Wittfogel, K. (1957). *Oriental despotism*. New Haven: Yale University Press.

Wolf, E. (1982). *Europe and the peoples without history*. Berkeley and Los Angeles: University of California Press.

Wrigley, E. (1969). *Population and history*. New York: McGraw-Hill.

Wunder, H. (1985). Peasant organization and class conflict in eastern and western

Germany. In T. Aston and C. Philpin, eds., *The Brenner debate: Agrarian class structure and economic development in pre-industrial Europe*. Cambridge, England: Cambridge University Press.

Yadava, B. (1966). Secular land grants of the post-Gupta period and some aspects of the growth of the feudal complex in northern India. In D. Sircar, ed., *Land system and feudalism in ancient India*. Calcutta, India.

———. (1974). Immobility and subjugation of Indian peasantry in early medieval complex. *Indian Historical Review* 1:18–74.

Yang Lien-sheng (1970). Government control of urban merchants in traditional China. *Tsinghua Journal of Chinese Studies* 8:186–206.

Yapa, L. (1980). Diffusion, development, and ecopolitical economy. In J. Agnew, ed., *Innovation research and public policy*. Syracuse, NY: Department of Geography, Syracuse University.

Zwernemann, J. (1983). *Culture history and African anthropology. Acta Universitatis Upsaliensis, Uppsala Studies in Cultural Anthropology*. Uppsala, Sweden: Uppsala University.

Index